WORLD ATLAS of GEOLOGY and MINERAL DEPOSITS

WORLD ATLAS of GEOLOGY and MINERAL DEPOSITS

DUNCAN R. DERRY

assisted by

Laurence Curtis   Soussan Marmont

Derek Fisher   Barbara Kwiecien

MINING JOURNAL BOOKS · LONDON

HALSTED PRESS DIVISION

JOHN WILEY & SONS · NEW YORK

First published in 1980
by
Mining Journal Books Ltd.
15 Wilson Street, London EC2M 2TR,
England

Map Drafting: Gervais Mineral Exploration Services
Atlas and Jacket Design: Victor Clark

Printed by Loxley Bros. Ltd.
Sheffield, England

Distributed by Halsted Press,
a Division of John Wiley & Sons, Inc., New York

Library of Congress Cataloguing in Publication Data
Derry, Duncan R.

A Concise World Atlas of Geology and Mineral Deposits
    1. Geology—maps
    2. Mines and mineral resources—maps
    I. Title
    G1046.C5D45 1980 912'.155 80-675233

ISBN 0-470-26996-0

CONTENTS

FOREWORD

Two great advances in thought about world geology can be discerned in the literature of the past three decades. On the theoretical side, the breakthrough in global tectonics represents a major advance towards James Hutton's ideal of a comprehensive theory of the Earth. The vindication from palaeomagnetic measurements of the earlier but, by many, neglected idea of continental drift; the establishment by bathymetric and geological survey of the ocean floor of the continuity of the mid-ocean ridge and rift system; the proof from stripes of normal and reverse magnetisation symmetrically arranged on either side of the ridge system that the ocean crust originates at the rift and spreads outward; the confirmation of this from movements of truncated volcanic cones and from radiometric dating; all these and other considerations have led to the concept that the earth is covered by a number of plates about 100 km thick, slowly moving under the influence of convection in the deeper mantle. The uneven distribution of fossil fuels and of the products of processes leading to the concentration of metalliferous minerals are illuminated and take on fresh significance in the light of this hypothesis.

On the practical side, the extraordinarily rapid pace of discovery and development of petroleum resources in the 1950's which prepared the way for the predominant position in world energy supply now occupied by oil and gas also in due course brought about a realisation of the precariously limited nature of these non-renewable resources. The attempt to evaluate world resources of workable coal and uranium ore has produced results that, seen against the rapid exponential rise in human population are hardly reassuring. We now know that world trade and global politics are dangerously subject to these factors. Resources of the other minerals that are basic to industrial society have also come into question, not in most cases because there is any discernible shortage of them in the Earth's crust, but because their recovery from lower-grade deposits would certainly be affected by the energy-intensive nature of the operations.

For both practical and theoretical purposes, the appearance of this World Atlas of Geology is both timely and promising. It is the first of its kind to take into account the impact of these two lines of thought. It is also the first to abandon the complex nomenclature of stratigraphy in favour of depicting the rocks grouped according to their absolute ages, as indicated by radiometric dating. No-one more suitable to compile it could have been found than Dr. Derry. A very successful explorer for, and developer of minerals, he has worked in all the continents save perhaps Antarctica, maintaining a keen and critical interest, from his Cambridge days on, in both the practice and the theory of the geological sciences. The Atlas will, I feel sure, have a wide appeal beyond the limits of professional geology. The accompanying text has been prepared without recourse to the jargon in which geologists converse and carry on their business.

The Atlas should be seen as a guide to the Earth's outer crusts, and the distribution of useful minerals in them. Inevitably some readers will require greater detail, and this is to be obtained primarily from the national geological surveys and allied institutions, a list of which is given. It is worth emphasising that these exist to provide, for public use, detailed geological maps, explanatory texts and information about national mineral resources. It could be added that even this information will never be complete until the last fuel reservoir or ore zone has been worked out.

It is a pleasure to commend this book to all those who wish to understand better the Earth we live on, and its natural wealth.

KINGSLEY DUNHAM

ACKNOWLEDGEMENTS

An undertaking of this nature involves reference to hundreds of maps, reports and books besides the innumerable conversations with people who have special knowledge or experience of the geology and mineral deposits of various parts of the world. Accordingly, it is impossible to quote all references and acknowledge all sources of information. However, certain people and organisations deserve particular thanks for their help in compiling this Atlas.

I was encouraged initially in proceeding with the Atlas by Sir Kingsley Dunham F.R.S., former Director of the Institute of Geological Sciences of the United Kingdom, and subsequently by the late Ian Campbell who was prominent in the formation of the American Geological Institute; one of the principal objectives of the Institute was the presentation of geoscience in non-technical terms. J. Tuzo Wilson and Hugh Wynne-Edwards were particularly helpful in making comments on some aspects of the early versions of the text.

D. J. McLaren, Director of the Geological Survey of Canada, and many of his staff, including J. O. Wheeler, W. W. Hutchison and the late R. J. W. Douglas, have been most helpful. E. Irving of the Earth Science Branch, Department of Energy, Mines & Resources, Canada, provided the base for two of the map projections used and gave advice on aspects of plate tectonics. At the Geological Department of the University of Toronto many people have given me the benefit of their interest and wide experience in the field of economic geology.

F. M. Vokes of the Geologisk Institut Norges Teckniske in Trondheim provided much information in regard to mineral deposits of northern Europe. I am indebted to Janet Watson of the Imperial College of Science and Technology, London, for advice and criticism on the European aspects of the Atlas and to A. R. Crawford of the University of Canterbury, Christchurch, New Zealand, for information on the complex tectonics of southern Asia and Indonesia. R. P. Hewitt and A. E. Marshall were also helpful

independently in reading and commenting on drafts of the Australian section.

Stephen Kesler of the University of Michigan provided his extensive knowledge of Central American ore deposits. Rex Birch brought me up-to-date on oil and gas in the North Sea and vicinity through his long association with the industry. Wilfred Walker gave me the benefit of his studies on global tectonics. Jason Morgan at Princeton gave valuable advice on some recent developments in plate tectonics and Al Fischer, also of Princeton, provided comment on the remarkably wide field of geoscience in which he is a respected authority. Ulrich Peterson of Harvard was most helpful on aspects of mineral distribution.

Global mineral production data was given by Gerry and Marilyn Govett, both in conversation and by extensive reference to their book. Paul Kavanagh made available his data on world gold production and Desmond Pretorius his work on gold and diamonds in South Africa.

Of the very numerous maps used in the preparation of this publication, the existing compilations of continents and large countries have, naturally, been utilised heavily. In this regard government and international organisations continue to make very valuable contributions to the earth sciences.

Data presented in the tables of production and reserves of the various minerals was derived mainly from *Mineral Trade Notes* and *Mineral Commodity Summaries*, both produced by the United States Bureau of Mines, with some modifications where more up-to-date figures could be obtained. The figures on coal, are from the publication *World Coal*, November 1979, and on oil and gas from the *Oil and Gas Journal*, December 1979.

A grant from the Canadian Geological Foundation in the early stages of compilation, advanced through the Canadian Geoscience Council, was particularly

encouraging at a time when the project might not otherwise have proceeded. I am especially grateful to the estate of the late Joseph S. Stauffer for a generous grant, also arranged through the Canadian Geological Foundation and Canadian Geoscience Council, that ensured the quality of the publication, most notably the production of negatives for the maps.

It would have been impossible to have completed this Atlas without the energy, skill and dedication over successive periods, of my associate authors, Laurence Curtis, Soussan Marmont, Derek Fisher and Barbara Kwiecien, and the patient and conscientious drafting skills of Donald Gervais and William Syrett and others of Gervais Mineral Exploration Services Ltd. I am also indebted to my partners and the secretarial staff of Derry, Michener & Booth in the diversion of time and facilities to the project on numerous occasions.

Marinus Kluyver was helpful in various ways in the early days of the project and of particular value in his careful reading of galley proofs and making constructive suggestions in the final review.

To my son, Ramsay Derry, I am grateful for comment and advice from the standpoint of a professional editor, unbiased by an earth science background.

Finally, I would like to acknowledge the forebearance of my wife Alice in accepting, for over four years, a cluttered study and a dining room table covered with maps on many weekends.

DUNCAN R. DERRY

INTRODUCTION

For various reasons, the interest in Earth Sciences by people outside that professional field has increased noticeably in recent years. Concern with air and water pollution on a global scale is one, the practical limits of world mineral resources, and their distribution by countries, another. We are being persuaded to look at the Earth as a unit, rather than as individual continents, much as we look at the Moon as a unit when we assemble the evidence collected on the lunar landings.

There are many excellent books on geology, generally and as applied to specific continents or countries. What seems to be unavailable is a publication consisting predominantly of geological maps and dealing with the earth as a whole, which is presented in a form that can be understood by the layman and yet can be useful to professional earth scientists, many of whom tend to have a limited geological knowledge of areas outside their field and areas of specialisation. A publication of this sort was produced about 14 years ago in German but nothing recently, or in English. This is understandable in that there are problems and restrictions in showing sufficient detail to be effective without exceeding what is practicable in bulk and cost, for the use of the Atlas by the student, the travelling amateur and the professional earth scientist.

Nevertheless, the present is a particularly suitable time to produce a compilation of this sort. It is no longer believed that the continents have always been in their present positions and the plate tectonic concept, involving the movement of continents, has been developed in more detail and has become accepted by the majority of earth scientists.

This concept assumes that the continents lie on a number of separate crustal plates that are continually moving and jostling each other; such movements, measurable in centimetres per year, include parting from each other, colliding, undercutting or grinding past each other. These movements obviously affect not only the geological history of the individual continents, including the formation of mountain systems, but also the distribution of life and the origin and distribution of mineral deposits. In this Atlas the main features of the plate tectonic concept are reviewed and, in the notes on each map sheet, the major stages of at least the later (post 250 million years—m.y.) history are related to plate movements.

With the above objectives and considerations in mind the following guidelines for this publication have been adopted:—

(1) Both maps and text should be comprehensible to those who have a minimal knowledge of geological science. Wherever possible complex terminology has been avoided

(2) The cost must be acceptable, at least in quantity orders, to reach a wide readership

(3) The dimensions and bulk of the publication must fit easily into a briefcase.

The first part of the Atlas, immediately following this introduction, is included in order to make the notes accompanying the map sheets intelligible to those readers who have relatively little background of geological science. There are many excellent books (some listed in the Suggested Reading chapter on page 102) that give this background much more comprehensively but these may not be available to the reader who is travelling or not near a library and some introductory outline of the composition and history of the Earth seems desirable if not essential. The professional earth scientist will find it largely superfluous.

The term "earth scientist" brings up the matter of terminology in the various branches of the science. "Geology" means the study of the earth and, in its original use, included all aspects whether of minerals, rocks, structures, fossil life or mineral deposits, including also the study and application of physical and chemical principles applying to the Earth's crust. Increasingly, however, especially in the mineral industry, the word "geologist" is being used in a narrower sense to exclude the "geophysicist" and "geochemist" specialising respectively in the application of physics and chemistry to mineral exploration. If an inclusive term is needed to take in the whole scientific family, the term "earth scientist" is logical and is used in this sense in the text that follows. A more detailed explanation of terminology is given in the Glossary which appears on page 108.

The second part of the Atlas consists of ten map sheets covering most of the land areas of the world. Each map is accompanied by a summary of the geological history of the particular part of the Earth's crust. In the introductory comments that precede the map section, some explanation is given of the reasons for variation in scale and projection from one map sheet to another and the treatment of the economic aspects.

One of the objectives of this Atlas is to present and discuss the very uneven distribution of mineral resources and the geological factors that control their genesis and location. In some cases the reasons for concentrations of a particular metal are still not satisfactorily explained. For instance, the fact that 55% of the known world resources of tin are contained in an area, centred on the Malaysian Peninsula, which is less than 1% of the surface of the earth. Other facts, such as that 50% of the world's copper production comes from a belt paralleling the west coasts of South and North America, can be accounted for within the plate tectonic concept reviewed on page 9.

This uneven distribution of mineral concentrations is of immense political and strategic significance and should be understood, within the limits of current knowledge, by everyone responsible for contributing to critical decisions affecting the supply and use of such resources.

LANDSCAPE AND GEOLOGY

The first thing a traveller notices about a new country is the existence and shape of such features as mountains, plains and escarpments. On a physiographic map, these may be shown by colour or by contours but this is not practicable on a map devoted mainly to showing geology. However, it does not take very much practice to get some idea of the sort of topography and landscape one can expect after studying a geological pattern on a map. Naturally, climate has a basic influence on scenery but, beyond this, the landscape depends upon the combination of geological processes. These include the surface processes such as erosion by water, ice or wind, and deposition by the same agents, and the folding and faulting of previously-formed layers together with the outpouring or intrusion of molten rock.

The surface agents of change can be sudden and cataclysmic, like avalanches or flash floods which can make changes to topography in hours or minutes. From these, in descending order of speed of action, one passes through the vigorous erosion of mountain streams and rivers, and the scouring of glaciers, to the more gradual changes produced by rivers in temperate climates and low elevations. Similar variation in the rates of wave erosion, but on a much smaller scale, occurs along the shores of seas and lakes.

The most fundamental influence on landscape, however, is the history of geological tectonic events, particularly folding, faulting, volcanic activity and intrusion. The crust of the Earth has predominantly a layered pattern of sedimentary and volcanic material that has, in places, been disrupted, contorted, and intruded by molten material, in zones controlled by broad crustal movements. The topography can tell us something of the type and degree of these disruptions or the absence of them. For example, a completely flat, extensive plain whether it be prairie, desert or tundra, is likely to indicate flat, undisturbed beds, at most very gently tilted. Examples on maps in this Atlas are seen in mid-western North America, in the Russian steppes and the western desert of Egypt.

Conversely, a broadly level surface, but with an uneven relief of low ridges and valleys (giving innumerable lakes in a glaciated and well-watered surface like northern Canada), is produced by erosion prevailing over hundreds of millions of years and wearing down old mountain systems to produce what is known as a *peneplain*. Examples are seen on the maps in the older parts of the Precambrian shields of Canada, Western Australia, Brazil and Zimbabwe.

A landscape with changes of scenery in roughly parallel belts, e.g. a belt of rolling hills succeeded by one of flat fields, and perhaps an escarpment similarly aligned, indicates gently-tilted sediments, such as in central and southeastern England.

A mountain range of approximately parallel ridges indicates a system of strongly-folded and overthrust sediments, possibly with volcanic layers, geologically fairly young—less than 500 m.y. Examples are the Himalayas and Alps (the most youthful being less than 50 m.y.) and the Rocky Mountains and the Andes (mostly younger than 200 m.y.). Rather less precipitous are the Appalachians of North America and the Caledonides of northern Europe which were at one time a single range and were formed between 500 and 400 million years ago. In general the younger the mountain range the more rugged and spectacular, i.e. less worn down.

A very rugged plateau, bordered by steep escarpments, often indicates an area of resistant lava, or an intrusive sill, overlying softer sediments. A spectacular example is seen in the southern Africa sheet in the Drakensberg Mountains of South Africa and the adjoining basalt-covered mountains of Lesotho.

The beautifully symmetrical volcanic cones such as Japan's Fuji Yama, New Zealand's Mount Egmont and the gentler-sloping Mauna Loa in Hawaii (if measured from the surrounding ocean floor, the world's highest mountain) are self-explanatory in origin. Less obvious is the origin of the rugged, spectacular scenery of many Pacific, and other oceanic islands which are also mainly volcanic. The irregularities in topography are due to the diversity of volcanic material (lavas, ash, boulder beds, etc.), each unit having a different resistance to erosion.

Most flat ocean islands, such as Bermuda, Bahamas and Barbados, have volcanic bases but are capped by coral reefs and sands.

Ocean Geology and Topography

It should be noted that the cumulative total of land area represents only 29% of the world's total surface. About 11% is occupied by continental seas, i.e. parts of the continents that are currently below sea level, while 60% is covered by oceans. Both continental seas and oceans have their own characteristic topography dependent on the geological framework. Most of the continental seas are included on the respective map sheets in this Atlas but it is not feasible to show, on a similar scale to the land masses, the oceans that form the larger part of the Earth's crust. The major structural features of the oceans, however, are shown on the small scale map in figure 3A (p10).

Continental seas lie on crust that has much the same geological character and composition as the land masses. Their topographical features, however, tend to have less sharp differences in elevation due to the lack of present vigorous erosion by, for example, mountain torrents and the rapid infilling of valleys by marine sedimentation.

Oceans, on the other hand, have a radically different geological make-up and correspondingly different physiography. It is only in the last twenty-five years, as a result of geophysical studies and the extensive programmes of deep drilling from specially equipped vessels, that we have learned much about the history and composition of the floors of the oceans. The first geological difference is that ocean floors have a composition that is characteristically that of basaltic lavas with only rather thin coverings of

organic or muddy material except under special conditions of strong currents or near continental slopes. Secondly, ocean floors are geologically young—in most cases less than 200 m.y. which is less than the most recent 5% of the Earth's history. Within this age range the older lavas tend to be near the margins of individual oceans while the younger, are generally near what are termed *mid-ocean ridges*. These are very extensive chains of sub-marine mountains and hills following long fracture systems up which basalts are being expelled. These are referred to later in the review of the plate tectonics concept.

In contrast to the mid-ocean ridges are the very deep trenches in the ocean floors, such as are shown in the Americas sheet of this Atlas, off the west coast of South America; in the Indonesia-Australasia sheet, south of the Indonesian islands and extending northerly from the east side of New Zealand; and in the Asian sheet, off the east side of the Philippines and the Japanese islands.

The most extreme topographic features mentioned above, both on land and under the oceans, are due predominantly to horizontal and vertical movements of parts of the Earth's crust. To understand the origins of these movements, and the principles involved, we need first to review the structure of the Earth from crust to core.

Evidence from measurements of the radio-activity of rocks and minerals on the crust of the Earth, supported by more recent data from lunar landings, show that the Earth was formed as a solid body about 4,600 million years ago. To support this figure and subsequent dates of significant events in Earth history, it may be worth reviewing briefly the methods used in dating rocks.

Rocks can be dated by four main methods:

(1) By studying and recording the succession of layers or strata—if the series has not been drastically disturbed this gives an age relative to a formation above or below on which we may have more detailed information.

(2) By identifying animal or plant fossils known to have existed for certain limited periods of time. This method, until 30 years or so ago the most widely applied to rocks younger than 570 million years (m.y.), gives a generalised age based on previously compiled evidence, but not a specific date in years. Obviously its direct use is limited to fossil-bearing sediments but igneous rocks may be dated indirectly by their interlayering with, or intrusion into, such sediments.

(3) By radioactivity—using the known rate of decay of certain isotopes such as uranium, potassium, rubidium or carbon. This method gives an absolute age, in years before the present, within the limits of laboratory accuracy. This is the most widely used method today and the ages plotted on these maps generally depend on it.

(4) By the magnetic orientation of certain minerals. It is a known fact that magnetic poles wander and reverse periodically from north to south. Magnetite, and other iron-bearing minerals, crystallise with their magnetic poles orientated in relation to the poles of the globe which existed at that time. The method only became feasible with the **development of plate tectonics because we must have some** idea of the position of the continent we are studying, relative to the geographic and magnetic poles of the earth, in the general period of the formation of the rock being measured. The method is useful especially in dating igneous rocks when radiometric methods are not applicable.

On the maps, and in the notes, in this Atlas the ages of rocks, in millions of years before the present are used as far as possible in preference to an extensive use of formational or time names. The dates plotted on maps have been generalised in many cases and do not include the margin of error customarily shown on detailed geological maps.

An unbracketed age date on a map refers to the original formation of the rock—for example, in the case of a sediment **the start of deposition of the system; of a lava, the date of its** solidification and of a granite, the time of its intrusion and solidification. A bracketed date on a map indicates the period of recrystallisation involved—for example, in a period of mountain building (orogeny). Such recrystallisation takes place in the course of metamorphism (from the

Figure 1: Geological Clock—This Spiral Chart Illustrates Graphically the Relative Duration of Geological Periods. Each Revolution Represents One Billion Years

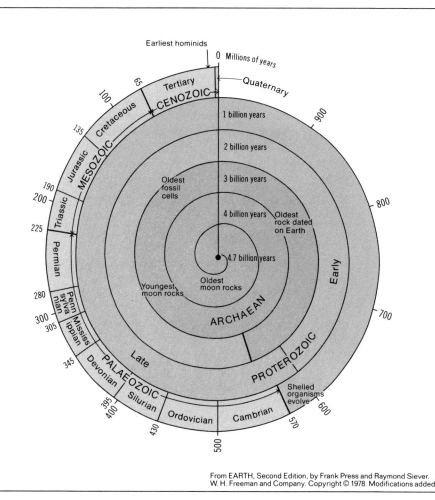

From EARTH, Second Edition, by Frank Press and Raymond Siever. W. H. Freeman and Company. Copyright © 1978. Modifications added.

Greek: *meta*—change and *morphos*—shape) which commonly accompanies mountain building and intrusion and this tends to "re-start" the radioactive clock for radiometric dating. The ages so obtained are those of such events rather than of the original deposition or crystallisation of the particular formation.

The relationship of dates in years before the present, to the major geological periods or systems is shown in the spiral "clock" in Figure 1. Most of these periods are named from geographical locations in Britain and northwest Europe where geology was first mapped. Thus "Cambrian", "Ordovician" and "Silurian" periods were named after the Roman names of tribes living in those parts of Wales that are underlain by rocks of these time divisions. "Devonian" is named after the county of Devon where rocks of this age are widely exposed. "Carboniferous" is the name given to the system that is coal-bearing in its upper part. "Permian" is named after the province of Perm in Russia and "Triassic" from the three-fold division of this system; "Jurassic" from the Jura mountains where rocks of this age were first described; and "Cretaceous" from the Latin *creta* (chalk) which is its most visible characteristic in England and France.

Figure 2: Internal Structure of the Earth

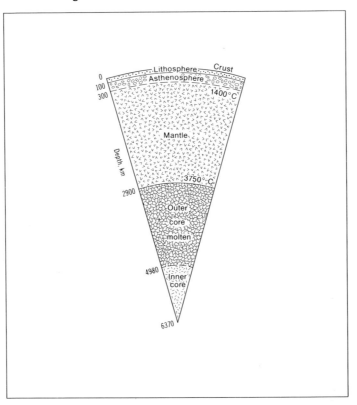

The term "period" applies to the time division, "system" to the series of rocks deposited in the corresponding period.

Prior to the wide use of radiometric dating, there was a tendency to subdivide formations in the younger part of the Earth's history in much more detail than the older. This resulted in students of that day spending much time memorising lists of formational names in the last 15% of Earth history with little regard to the previous 85%. The present attitude, thanks mainly to radiometric dating, is much less lopsided. To keep our sense of time in perspective we should remember that the period that has elapsed since the earliest direct ancestors of the human race were living in the Great Rift Valley of Africa—about five million years ago—is slightly over one-tenth of 1% of the total history of the planet.

Structure of the Earth

It has been known for over a century that the central part of the Earth consists of a core of white-hot fluid material denser than the outer shell, called the mantle, which in turn is denser than the outermost layer, the crust. (Figure 2.)

In simplified terms, a section through the solid Earth consists of:

(1) The *crust* consisting of two types of material—
(a) The oceanic crust of dense basalt lavas which directly underlies the oceans and averages about 7km depth;
(b) The continental crust, consisting mainly of granitic and sedimentary rocks which form the continental masses, is of lower density and averages 30 to 40km thickness increasing to as much as 70km under some mountain ranges such as the Andes.

(2) The *mantle,* underlying the crust. This seems an unfortunate term giving to the uninitiated a picture of something covering the whole Earth like a blanket. It does, of course, cover the core but its upper level is anything from 5 to 70km below the Earth's surface. The mantle is denser and hotter than the crust. It is solid but a layer from, perhaps, 150 to 400km below surface is hot enough to be slightly plastic and yielding. At greater depths the pressure makes the deeper mantle rigid again.

(3) The *core* extends from the boundary with the mantle, at about 2,900km depth from surface, to the centre of the Earth at 6,370km. It appears to be divided into a solid *inner core* with a radius of about 1,400km and a liquid *outer core* which is 2,080km thick between the inner core and the

mantle. The outer core has a lower melting temperature than the mantle.

We have no positive information on the composition of the core but speculation based on density and seismic measurements suggests that it is composed primarily of iron and nickel in combination with a lighter element such as silicon or sulphur.

Migration of Continents

When we think of geological processes affecting the crust of the Earth below surface we must realise that although rock behaves like a hard, brittle substance when struck with a hammer or drilled, it will yield as a plastic material under long, continued pressure, particularly under temperatures higher than those prevailing at the surface. This feature explains the often complex folding of volcanic and consolidated sedimentary rocks and the vertical movements, on a large scale, of parts of the continental crust. In the latter case, if an extra weight forms on part of a continent, an ice cap for example, then that part will sink deeper into the yielding, denser material on which it is "floating" like an unevenly-loaded raft. Similarly, when a mountain range is worn down by the action of glaciers and rivers carrying eroded material into the sea (i.e. lightening the weight of the land), the part of the continent under the eroded mountains will rise like an unloaded raft.

The same capacity of rocks to yield under increased temperatures permits the horizontal movement of segments of the crust. Actually it is now found, through the study of earthquake waves, that the most critical change from relative rigidity to plasticity is not at the crust/mantle boundary but within the mantle at about 150km depth. This is the boundary between what is known as the "lithosphere" (from the Greek: *lithos*—rock) which includes all the crust, whether oceanic or continental, and the upper part of the mantle to about the 150km depth. The plastic and more yielding layer of the mantle below this depth, down to somewhere between 400 and 700km has been named the "asthenosphere" (a clumsy word but legitimately begotten from the Greek: *asthenos*—weak). The concept of plate tectonics, depends on this yielding layer for the horizontal movement of plates and for slow convection currents below the plates that may instigate crustal migration.

Until the 1950's it had been accepted by most earth scientists that the continents had always had approximately the same position and shape that they now have. Although

some people had noted the "jigsaw" effect in which, for example, the west coast of Africa "fits" into South America, it was the German scientist, Alfred Wegener, who made the specific proposal in 1912 that over 200 million years ago the present continents (including their sea-covered continental shelves) were packed together in one giant landmass which he termed "Pangaea". This supercontinent was semi-divided into two portions which have subsequently been called "Laurasia" (including North America, Europe, Asia and the Arctic), and "Gondwana" (which was named earlier by the Swiss geologist Edward Suess and included South America, Africa, India, Australasia and Antarctica). This **Pangaea landmass**, according to Wegener's concept, had split apart in stages with continental masses drifting off like floating ice in the sea. He pointed to similarities in geology, animal and plant life, and also in former climates, between the now widely spaced continents.

Wegener's ideas, although inspiring considerable interest, received little support and some vitriolic opposition during the next twenty years. Exceptions to the detractors included Arthur Holmes of Edinburgh and A. L. de Toit of South Africa, both of whom supported, in different ways, the essential Wegener concept. Wegener died on a trip across the Greenland ice-cap in 1930, his ideas still unaccepted by the majority of earth scientists. It was not until the mid-1950's that the concept of "continental drift", and its successor "plate tectonics", received fresh and compelling support and, by 1970, continental movement by plate tectonics was accepted by the majority of earth scientists.

The Plate Tectonic Concept

A combination of factors led to what is unquestionably the greatest revolution in earth science since its inception. This included the increased availability and efficiency of air travel which enabled a large number of earth scientists to study and compare geological evidence in widely spaced parts of the world; advances in radioactive dating of rocks on a world-wide basis; the widespread measurements and understanding of magnetic variations over the world; and technological advances in the drilling of holes into the floor of deep parts of oceans.

The first observations that led to the revival of Wegener's theory, and its modification to the "plate tectonic" concept, were in the mid-Atlantic Ridge. This extensive north-south structure had been known from soundings for many years as an "ocean mountain range" but its extent and significance had not been recognised. Since the Ridge, which rises to heights of 3,000 and 6,000 metres above the surrounding ocean floor, was found to consist largely of volcanic material, it was suggested (originally by H. H. Hess) that this might be the line of parting between the two sides of the Atlantic Ocean with new volcanic material rising to fill the gap created by the parting ocean crust.

The evidence that confirmed this concept of a line of parting being filled by successive outpourings of lava came from two main sources. The first is what has been called the "magnetic signature" in rocks. It had been known for some time that the magnetic poles have "wandered" throughout geological time and that there have been periodic reversals of the north and south magnetic poles. When molten material, like lava, crystallises the magnetic materials are orientated in a position corresponding to the Earth's magnetic poles at that time and are frozen in this position. Thus, when such rock is examined it is possible to tell the

position of the magnetic poles, and whether they were "normal" or "reversed", at the time of solidification of the rock and thus the time this occurred. The second source of evidence was from a succession of drill holes into the ocean floor, particularly from a specially equipped vessel the *Glomar Challenger* which has enabled numerous and widespread samples to be collected and dated by radiometric methods.

On the basis of both techniques, it was found that the lavas closest to the parting fissure were the youngest and flows on each side increased in age with distance, each side matching the other as a "mirror image". There seemed little doubt, therefore, that "ocean spreading" was taking place in the Atlantic. The two sides of the ocean had progressively parted over the past 180 million years and were continuing to diverge at a rate of two to four centimetres a year. The Ridge system of the mid-Atlantic (Figure 3A) was found to

Figure 3A: Plates and Plate Boundaries of the World

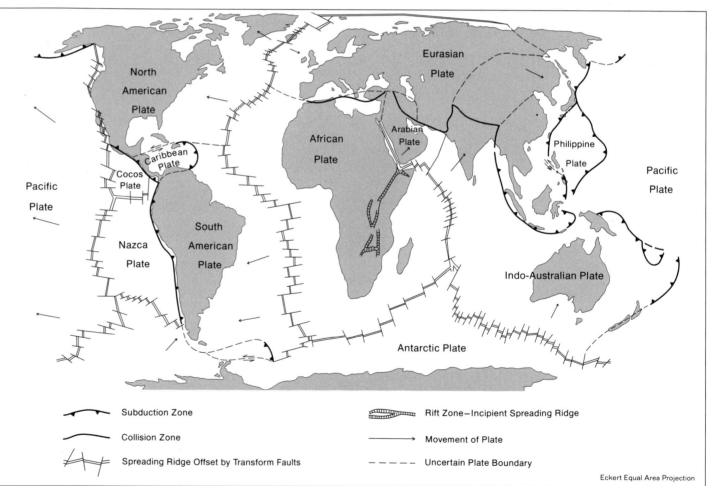

be traceable all the way from the north of Iceland, down through the Atlantic Ocean, around the foot of Africa and into the Indian Ocean where it splits into two branches. One branch goes into the Red Sea while the other passes between Australia and Antarctica into the Pacific (where the rate of spreading reaches as much as twelve centimetres a year) and runs into the west side of North America in the region of the Gulf of California.

The conclusion that "sea-floor spreading" exists led to the concept that, instead of Wegener's drifting of continents through the denser oceanic crust, they may form integral parts of lithosphere plates and be carried on them, "similar to logs frozen into ice flows on a lake surface", as Peter Wyllie puts it. New lithosphere is being generated at the ocean ridges pushing apart older lithosphere which is being carried away on each side as if on conveyor belts moving in opposite directions. This process, however, brings a new

problem if it is assumed that the diameter of the Earth is more or less constant, as is believed by a majority, but not all, of earth scientists. If the sides are parting at a fissure in a mid-ocean ridge, the surface material must go somewhere because of the necessity of shortening the crust to balance the ocean spreading. The shortening is too much to be accounted for by the crumpling of the surface in mountain building (Figure 4).

The best answer lies in the concept of "subduction" which, expressed simply, means that when a moving plate meets another that is static, or moving in the opposite direction, one of the plates underthrusts the other and, as it descends into the hotter interior, melts and becomes assimilated into the surrounding mantle material. If it is oceanic crust that meets continental, it is the former that underthrusts the latter on a line paralleling the plate boundary. Along such a subduction zone (also called a "Benioff zone") the first in-

Figure 3B: Precambrian Basement and Foldbelts of the World

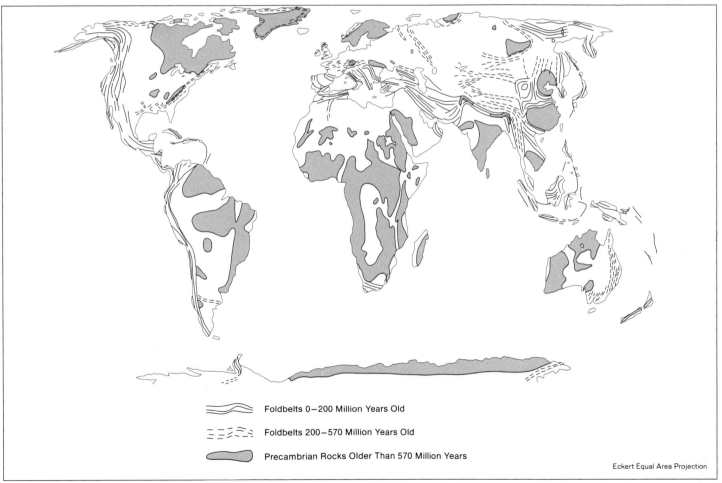

~~~ Foldbelts 0—200 Million Years Old

- - - Foldbelts 200—570 Million Years Old

▬ Precambrian Rocks Older Than 570 Million Years

Eckert Equal Area Projection

dication at surface is the formation of a deep ocean trench. Examples of these may be seen in the maps in this Atlas such as off the west coast of South America, the southern coast of the Indonesian islands, the east coast of New Zealand and extending northerly and on the ocean side of the Japanese islands. The sinking trenches accumulate great thicknesses of sedimentary material derived from the erosion of the continent, often accompanied by volcanics, and as the two plates continue to converge this stratified succession undergoes crumpling and thrusting along a line approximately parallel to the subduction zone. This deformation at the continental margin is known as an orogeny.

Orogenies are mountain building movements involving folding and faulting. Nearly always they are accompanied by volcanic activity and granitic intrusion which are believed to result from the melting and upward migration of subducted material from the underthrust plate. All the major mountain systems which have developed in the past 250 million years are definitely related to plate collisions, e.g. the Cordillera of the Americas, the Alps, the Himalayas. The relationship of these mountain systems to plate boundaries is seen in Figures 3A & 3B.

A feature that frequently occurs adjacent and parallel to a subduction zone in an ocean is the formation of "volcanic island arcs" between the subduction zone and the edge of the continent. Island arcs are elongated arcuate ridges commonly showing at the surface as chains of volcanic islands. Those that we see today are related to mountain building that is still active, or at least geologically recent, and are probably mountain belts in the process of formation. Examples are the Japanese islands, the Philippines and the West Indies.

We have defined two types of structures that form plate boundaries: the "diverging" ones where new crust is created and thus have also been termed "constructive" boundaries (for example, the mid-Atlantic Ridge or the geologically-recent fracture zone along the Red Sea, seen on the North Africa sheet); and the "converging" boundaries at subduction zones where crust is destroyed and these have been termed "destructive" boundaries.

A third type of boundary is where one plate slides laterally past another. This type has been termed a "conservative" boundary by some writers to denote the fact that on such structures crust is neither created nor destroyed. Such boundaries are characterised by having a high concentration of earthquake centres but little volcanic activity.

In Figure 3A, where diverging plate boundaries are shown, it may be noted that the central fracture, or plate boundary, has been offset laterally on numerous smaller fractures nearly at right angles to it. These are known as "transform faults" (originally defined by J. T. Wilson) and they allow for the adjustment on plate boundaries that is necessary when rigid plates move over the curved surface of the earth. In this connection it must be realised that plates rarely move directly away from or towards each other since there is nearly always some rotation in the plate movement. Also in Figure 3A it may be seen that a transform fault on which there is significant lateral movement itself becomes a plate boundary of the "conservative" type.

Transform faults also occur along subduction or collision zones but usually form sharper angles with them than they do with diverging plate boundaries. Such transform faults, forming branches of "destructive" or "conservative" plate boundaries, are particularly important where there is strong lateral movement. An example is seen in the American sheet of this Atlas in the northerly to westerly movement, amounting to several hundred kilometres, of the Pacific plate relative to the American plate along the plate boundary that approximately follows the Pacific coast of Alaska. Another example is the large lateral movement on the Alpine Fault of New Zealand where the Pacific plate has moved, and is still moving, northerly relative to the Indo-Australian plate.

Figure 4: Plate Tectonics: A Cross-sectional Representation and Summary

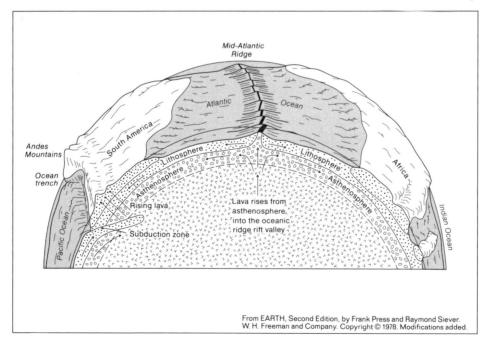

From EARTH, Second Edition, by Frank Press and Raymond Siever.
W. H. Freeman and Company. Copyright © 1978. Modifications added.

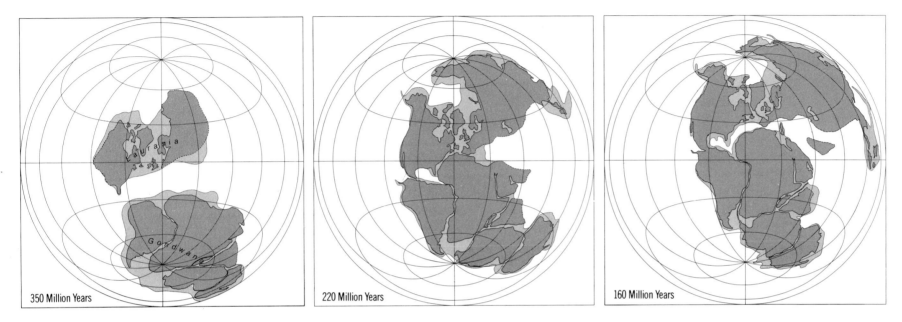

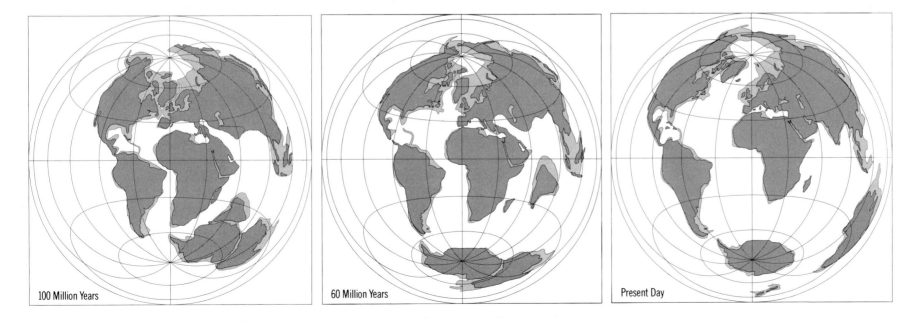

Figure 5: The stages of global tectonic history from 350 m.y. to the present. They are plotted on Lambert's equal-area projection which gives a global view and preserves the relative differences among its areas but includes over 75% of the total surface of the globe. Latitudes and Longitudes are marked at 30° intervals about a centre of Greenwich Longitude at the equator; thus the extreme Longitudes shown are 150°E and 150°W.

Today's coastal outlines are shown simply to identify areas and bear little relationship to the actual coastlines at the particular times. On the other hand, the continental margins are thought to have a closer resemblance to those that existed at the respective times.

The continental positions represent a consensus of recently published information especially that by A. G. Smith and J. C. Briden, E. Irving and R. H. Sillitoe.

It was noted earlier that there is substantial agreement that the present continents were grouped together in the period prior to 200 m.y. and commenced parting after that date by the movement of plates. Today, there are six major plates (the Eurasian, African-Arabian, American, Indo-Australian, Pacific and Antarctic) and a number of smaller ones. Some stages of plate movement since 220 m.y., and one stage before this, are shown in Figure 5.

To summarise the present (1980) ideas involved in the plate tectonic concept:

(1) The Earth's lithosphere (crust and upper mantle) to a depth of about 75km has significant strength, in the sense of possessing resistance to earthquake shock waves, and is relatively rigid.

(2) The lithosphere overlies a zone of weaker and hotter rocks termed the asthenosphere. It is at or below this boundary that the upper part of the Earth moves relative to the core.

(3) The surface of the Earth consists now of six major plates and a number of smaller ones. These are rigid but move relative to one another, the plane of movement being at a depth of about 150km where the change from relatively rigid to plastic material occurs.

(4) The boundaries of the plates, along which movement takes place, are of three types:-

(i) Diverging (constructive) where new crust is created by upwelling of volcanic material from the mantle, resulting in ocean-spreading and marked by ocean ridges such as that in the mid-Atlantic.

(ii) Converging (destructive) where one portion of the crust (commonly oceanic crust) is consumed by being forced beneath another, commonly a continental mass. This process is known as "subduction" and the zones, usually curved in plane, on which this process occurs, localise the formation of mountain ranges. In the oceans, the main zones of subduction are often marked by parallel arcs of largely volcanic islands between the subduction zone and the continental mass, e.g. the Philippines and Japan. In the rarer instance, where two continental masses converge, their collision may result in massive orogenic pulses such as created the magnificent Himalyan and Alpine chains.

(iii) Horizontal sliding (conservative) where crust is neither created nor destroyed, one plate moving laterally past another.

(5) Adjustment during plate movement takes place on "transform faults" which tend to be orientated at high angles to diverging plate boundaries and at lower angles to converging ones.

(6) Two plates moving apart from a diverging boundary may be mainly oceanic crust but either or both may carry continents on them. At the opposite side of a diverging plate (where it meets another one) subduction (underthrusting) of one plate by the other may take place at the junction. If the plate carries, near its subducting edge, continental crust, this being lighter, tends not to be drawn down as much as oceanic crust but instead becomes involved in crumpling and folding parallel to the subduction zone. This crumpling also affects the sediments and some volcanics which have accumulated in the deepening trench that forms along the subduction zone during the early period in its development.

(7) Such mountain building process, or orogeny, at a plate boundary includes the uprising, as lavas and intrusions, of molten material derived from the melting of the crust of the subducted plate.

(8) All major mountain systems younger than 250 m.y. follow known converging plate boundaries and older systems are believed to have resulted from similar collisions.

DISTRIBUTION OF EARTHQUAKES AND VOLCANOES

The review of the processes involved in the plate tectonic concept leads logically to reasons for the concentration of earthquakes and volcanic activity along certain linear zones in the Earth's crust. Although the two tend to overlap this is not necessarily the case and the two phenomena are dealt with separately.

Earthquakes

A glance at Figure 6, showing the distribution of earthquake epicentres, makes clear the heavy concentration along or near plate boundaries. This does not mean that earthquake movements are necessarily exactly on plate boundaries, but they are likely to be on them or on faults that angle into them. For example, there is some question as to whether the San Andreas Fault through San Francisco, on which a west-side-north movement of 260 km has been shown to have occurred in the past 25 million years is the actual boundary between the Pacific and the American plates or whether it is a transform fault running into the main plate boundary lying further out in the ocean.

As is explained in any work on elementary geology, a fault on which movement may take place resulting in an earthquake may be any one of three classes:

(1) A normal fault, which is caused by horizontal tension, resulting in the overhanging side of an inclined fault (or either side of a vertical one) sliding downwards.

(2) A reverse fault, resulting in compression, in which the upper side of an inclined fault-plane moves "upslope".

(3) A lateral fault (also called "transcurrent" when of major proportions), in which one side moves horizontally relative to the other. (The movement on a fault of this type is described from the point of view of a person standing on one side of the fault and looking across it. If the far side has moved to the right, the fault is described as being "right-handed" or "dextral"; if to the left, "left-handed" or

"sinistral"). Naturally, such lateral movement may be combined with either normal or reverse movement.

Movements along any of these three types of faults may result in an earthquake of some sort but not necessarily catastrophic. Smooth, steady movement tends to give relatively little shock. Greater shocks result from the tendency to move being inhibited by an irregularity in the plane of fault movement causing pressure to be built up and then suddenly released as the obstruction is overcome. Lateral faults always affect the surface of the Earth directly while normal or reverse faults may often be deeply buried and the actual movement may not reach the surface. Consequently many of the most disastrous earthquakes in human history—such as those in China, in Shensi Province in 1556, at Kansu in 1920, and at T'angshan in 1976, in all of which fatalities were numbered in the hundreds of thousands—are believed to have been caused by lateral movement on extensive faults.

Volcanoes

Volcanoes are also concentrated more along plate boundaries than on other parts of the Earth's surface (Figure 6) although occurrences away from such structures are more common than in the case of earthquake epicentres. The eruption of a volcano is usually periodic, separated by dormant intervals, and may come from a fissure or a pipe. The resultant outpourings consist of varying proportions of lava, volcanic debris and volcanic gases. The volcanic vents or fissures may be submarine or sub-aerial, i.e. on land and in this case the vulcanism may be accompanied by extensive ash falls, gaseous explosions and volcanic ejecta. The ensuing catastrophic destruction has been recorded throughout history.

The character of a volcano is controlled largely by the composition of its lava or fragmental derivatives. In general if the lava is high in silica and low in iron and magnesium ("acid" in geological parlance) it is apt to be more explosive and thus produces a large proportion of fragmented material relative to flowing lava resulting in a sharp conical vent (e.g. Mt. Vesuvius). A volcano extruding "basic" lava, relatively low in silica and high in iron and magnesium, tends to be more fluid and less explosive and builds broad "shield" volcanoes with gentle slopes. For example, the basaltic volcanoes of Hawaii which have erupted frequently during the history of man; in fact, the most easterly and largest island is almost continuously active, but human casualties have been extremely rare. At the other end of the scale is the example of Krakatoa, an island off the southeast end of Sumatra in Indonesia, which has had the greatest volcanic explosion in recent history and where the extruded material was relatively high in silica and viscous resulting in a large proportion being released as fragmented explosion breccia and ash. Naturally, this type is likely to be more catastrophic to human settlements.

By far the greatest volume of volcanic material on the Earth's surface is that occupying the floors of oceans. As explained earlier this volcanic material is believed to have been extruded from plate boundaries of the mid-ocean ridge type as pairs of plates spread apart. Nearly all the material forming the floors of present oceans is younger than 200 m.y. In addition to the basaltic lavas forming the greater part of the ocean floors, island arcs—consisting of volcanic material related to subduction zone plate boundaries—constitute separate zones of volcanic activity. The lava is of a different composition to that forming the main bulk of the ocean floors and is usually extruded from pipes rather than from fissures. Japan, the eastern Caribbean Islands, and the Philippines are examples of island arcs.

On land the greater proportion of geologically recent volcanic material has been extruded close to plate boundaries and falls into two main classes:

(1) Volcanic activity associated with mountain building along plate boundaries of the subduction zone type. An outstanding example is the chain of volcanic areas formed during many successive periods along the whole length of the Cordillera in South and North America.

(2) Flood basalts (also called "plateau basalts") which have occurred at several periods in geological history. In such an event, a vast area is covered by successive outpourings of free-flowing basalt, mainly from fissures. Examples that are shown in the Atlas are the Deccan Traps of India, the Parana Basalts of South America, the Columbia River Plateau of USA, the Drakensberg Basalts of South

Figure 6: Plate Boundaries and Active Zones of the Earth's Crust

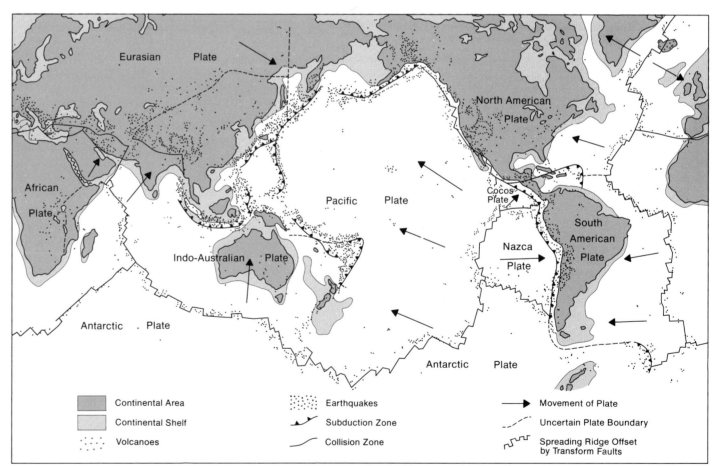

Africa and the Siberian Traps of Russia. In most of these cases, the release of lava has been related to rift systems, i.e. zones of incipient partition of continents. The most **geologically recent example** is the belt of basaltic lavas that follows the trend of the Great Rift Valley of East Africa.

A volcanic phenomenon that is an exception to the general rules of volcano distribution, and which has been recognised comparatively recently, is that described under the commendably simple term *Hot Spots*. These are isolated centres of periodic volcanic activity independent of plate boundaries, although a number of them do occur on such structures, and may even initiate them. Hot Spots appear to remain nearly stationary relative to the Earth's core, or at least to move more slowly than the plates which may move over them. By 1976, over a hundred Hot Spots, active in the last ten million years, had been recognised on the surface of the Earth with a rather uneven distribution. About 55% of these are on continents, with Africa having more than its share relative to area, and the remainder in ocean areas. Hot Spots represent less than one percent of the world's volcanic activity and the composition of the lava, basalt with an above-average content of alkali metals (sodium, potassium, etc.) differs from that of either the basalts that form ocean floors or the andesites and more silicic types that occur in continental mountain ranges.

In an ocean area, away from a plate boundary, Hot Spots may be recognised by groups of volcanic islands often forming straight or bent lines. In such a group the island nearest the Hot Spot is the youngest, as dated by radioactivity, and the island farthest away the oldest. The explanation is that, while the Hot Spot remains relatively stationary, the plate lying over it is in continual motion and each periodic outburst of activity forms a separate island volcanic centre. One of the best known examples is the Hawaiian group of islands where the largest and most easterly island, surrounding the huge cone of Mauna Loa, is the youngest (and still active) while the most westerly gives the oldest date. This pattern is caused by the westerly movement of the Pacific plate over the Hot Spot.

Iceland is believed to be an example of a Hot Spot on a divergent plate boundary. It is a 400 km diameter dome of volcanic material situated on the Mid-Atlantic Ridge (see Arctic sheet p90). The ridge itself has remained constant in position relative to the two adjoining plates and so the successive periods of volcanic activity have piled layer upon layer of lavas around the same general centre.

On a continent, a Hot Spot that lies on or near a belt of volcanic activity related to present plate boundary is difficult to recognise and there are probably many that have not been identified. A continental Hot Spot distant from a plate boundary, is more easily recognised and a particularly clear example is in Africa at Tibesti in the Sahara of Chad (see Northern Africa sheet p56). Since the African plate is believed not to have moved significantly in the last 30 m.y., this Hot Spot has resulted in successive outpourings of lava in the same centre forming a dome of over 200 km diameter.

One that has been recently recognised is the crescent-shaped area of volcanic rocks extending westerly from Yellowstone Park, Wyoming. Here radioactive datings show an increasing age from the current activity (geysers, etc.) in the Park to 15 m.y. in lavas 600 km to the west.

Hot Spots are believed to be caused by "plumes" of heat rising from the mantle through the crust and are thus related to the convection currents active within the asthenosphere. There is some evidence that Hot Spots may initiate the fissures that become divergent plate boundaries, i.e. may be the first stage in the break-up of continental masses.

LIFE

The history and dating of geological events must be studied in relation to the evolution of life and this aspect, and the use of fossils in identifying formations from one part of the Earth to another, has been a part of earth science since it was initiated. In the notes to a condensed atlas of this sort, the subject of life on Earth can barely be touched on and there are numerous excellent books that deal with the subject from the needs of the amateur to the professional. The following brief notes merely review the more important developments of life in relation to the major time divisions of Earth history. (See Figure 7.)

Not many years ago recorded evidence of life on Earth was restricted to fossil remains, or impressions, that could be seen without the aid of a microscope. More recently, with this aid, evidence of life has been extended back to 3,200 m.y. and may go beyond this. The forms of life at this time were those consisting of cells without nuclei, such as is the case today with blue-green algae and bacteria. At present, the earliest evidence is from the Fig Tree Formation near the boundary between South Africa and Swaziland (see southern African sheet p61) in rocks dated 3,200 m.y. Similar forms of life are known from somewhat younger Precambrian formations in North America and Australia. The oldest signs of life that can be seen without a microscope are structures known as "stromatolites", a rather inclusive term covering layers of algae separated by mud or lime resulting in lens- or pillow-shaped structures with concentric patterns. These have been known for some years in rocks of 2,000 m.y. or older on most continents but recently have been identified in Natal, South Africa, in rocks as old as 3,000 m.y., i.e. nearly as old as the oldest microscopic forms of life.

The development and diversification of life seems to have been very slow during most of the Precambrian era, due, possibly, to the very fact that the forms of life existing then were non-nucleated cells, i.e. the genes in each cell were not concentrated into a nucleus. Such cells are believed to be less susceptible than nucleated cells to mutation and thus to the evolution of new species. The low proportion of oxygen in the atmosphere in the early history of the Earth may also have inhibited evolution to more advanced life forms.

The first multicellular (and nucleated) organisms are found in rocks of about 700 m.y. i.e. near the end of the Precambrian era. Once this stage was reached the diversification of life advanced dramatically and, by the beginning of the Palaeozoic era, about 570 m.y., literally thousands of quite complex species had evolved of which those with a hard protective shell such as trilobites and shelled brachiopods were easily preserved as fossils. The rather mysterious stage of expansion in the preceding period between 700 and 570 m.y. probably involved life forms mostly without hard parts and, therefore, less easily preserved as fossils.

From 570 m.y., life expanded rapidly but was initially restricted to primitive water plants and invertebrate animals living in water. It is in rocks 480 m.y. old that the earliest fossil animals with backbones are found, these vertebrates being primitive fishes that led to the shark family.

The Devonian period, extending from 395 to 345 m.y. was a time of expansion of invertebrates, especially colonial corals which formed the extensive reefs that are the loci of many of the petroleum reservoirs worked today in North America and the USSR. Petroleum began to be generated with the expansion of life at the start of the Palaeozoic era but the Devonian period is the first that is seriously productive on a world-wide basis. In the same period, fishes became very abundant and diverse and, near the end, amphibians began to develop. With the spread of early land plants, insects first appeared.

In the following period (Carboniferous in Europe, Mississippian and Pennsylvanian in North America), from 345 to 280 m.y. the first true amphibians invaded the land. At least in lower-lying areas and warm climates, the land became increasingly covered by tree-ferns and other plants in sufficient abundance to be the source of coal beds. The variety of insects and spiders expanded with the growth of forests and the earliest reptiles appeared. In the seas, corals and sponges increased in numbers and the twin-shelled clams, or pelecypods, began to outnumber the assymetrically-shelled brachiopods.

The Permian period, from 280 to 225 m.y., brought to a close the Palaeozoic era and was characterised by an arid climate over quite a large proportion of the world. Life in the seas was reduced and many species of both vertebrates and invertebrates became extinct, as did many plants. On the other hand, the conditions favoured the development of reptiles on land. A branch of these evolved into the dinosaurs and, in the following period, this remarkable group began to occupy dominant positions in the life of land and sea.

TIME DIVISIONS			Age in Million Years	FOLDING–FAULTING–INTRUSION / FLOOD BASALT EVENTS / ICE AGES						LIFE		Age in Million Years
				America	Europe	U.S.S.R.	S. Asia	Australasia	Africa	Animals	Plants	
PHANEROZOIC — CENOZOIC — Quaternary		RECENT	0.2	Ice Age	Ice Age	Ice Age		Ice Age S. Latitudes	Rift Valley and Arabian Basalts			0.2
		PLEISTOCENE	2					Indonesia Papua–N. Guinea N.Z.				2
	TERTIARY	PLIOCENE	11	Columbia River Basalts	Alpine		Himalayan			First hominids 5 m.y. ±		11
		MIOCENE	25		Scotland–Ireland Basalts	Alpine						25
		OLIGOCENE	40							Expansion of mammals and birds	Expansion of grasses	40
		EOCENE	55				Collision of India with Asia 45 m.y.		Atlas folding	Major evolutionary change		55
		PALAEOCENE	65				Deccan Traps India			End of dinosaurs		65
PHANEROZOIC — MESOZOIC		CRETACEOUS	135	Cordilleran folding / Parana Basalts S. America	Pre-Alpine	Cimmerian	E. China granites	New Zealand Alps	Drakensberg Basalts	Peak of dinosaurs	Flowering plants (coal forming)	135
		JURASSIC	180		Hercynian	Siberian Traps			Cape folding	First birds	Conifers	180
		TRIASSIC	225			Southern Hercynian				Early dinosaurs / First mammals 210 m.y. ± / Major evolutionary change		225
PHANEROZOIC — PALAEOZOIC		PERMIAN	270	Ice Age in southern S. America			Ice Age India	Ice Age in south	Ice Age in south	First reptiles / Land insects	Abundant tree ferns— coal forming	270
		Upper Carboniferous (Pennsylvanian)	305			Urals folding	Ice Age	Ice Age	Ice Age			305
		Lower Carboniferous (Mississippian)	350	Appalachian folding	Caledonian folding			Tasman folding				350
		DEVONIAN	395							First amphibians	Early land plants	395
		SILURIAN	440							First fishes		440
		ORDOVICIAN	500					Adelaidian	Ice Age in Sahara			500
		CAMBRIAN	570						Pan African event	Expansion of shelled invertebrates	Sea plants	570
PRECAMBRIAN — PROTEROZOIC		LATE	1000	Ice Age / Grenville		Ice Age Baikalian	Ice Age in China			Nucleated cell organisms appear		1000
		MIDDLE	1800				Southeast India					1800
		EARLY	2600	Hudsonian Ice Age 2000 m.y. N. America								2600
				Superior								
PRECAMBRIAN — ARCHAEAN		LATE	3300						Zimbabwe	First signs of life (non-nucleated cells, e.g. blue green algae) about 3200 m.y.		3300
		EARLY	4500									4500

Figure 7: Time Chart of Geological and Life Events
(Vertical Scale not Proportional to Elapsed Time)

In this table which (unlike the spiral diagram in Figure 1) is not drawn proportionately to time intervals, ages in millions of years are related to the names of geological eras and periods. The relative magnitude of events involving folding, faulting and intrusion is indicated by the amplitude of the marked wavy lines, within vertical columns that correspond roughly with areas covered by the individual map sheets used in the Atlas. Periods of extensive flood basalt outpourings are marked by black columns and ice ages by these words.

In the right hand columns, the approximate stages of animal and plant evolution have been indicated to show their relationship to physical events and to time before the present.

It has been pointed out by several writers (recently by Dr A. G. Fischer of Princeton University in the 1980 William Smith Lecture to the Geological Society of London) that major geological events tend to show cyclical patterns. The most extensive cycles have a duration of about 300 million years. Each such cycle is characterised by the alternation of (1) a period of high sea level, low continents, low temperature variations latitudinally and vertically and generally stable conditions, and (2) a period of low sea level, high continents and relatively sharp temperature changes latitudinally and vertically, with a prevalence of ice ages in polar regions corresponding to the elevated continents. Two of these 300 million year cycles have occurred since the latest stages of the Precambrian era and are suggested in this chart by the wide distribution of ice ages at about 600 m.y., 300 m.y. and in the last two million years.

Within the 300 million year cycles there exist smaller cycles of about 30 million years.

The Triassic, beginning about 225 m.y. was the first period of the Mesozoic era and life in the seas began to increase again with many new types, including the spiral-shelled ammonites and a great increase in clams. But the most outstanding development in life in this era was the expansion in numbers and diversity of dinosaurs which reached their "golden age" from about 160 to 80 m.y., running or crawling on land, swimming in seas and lakes and flying in the air. In addition to the flying reptiles without feathers, such as the pterodactyl, the first feathered (and toothed) birds appeared about 150 m.y. ago. By the end of the Cretaceous, 65 m.y., the dinosaurs had become almost extinct and mammals, which had appeared in primitive form at about 200 m.y. expanded rapidly and took over the dominant position that had been occupied by the dinosaurs. Flowering plants had appeared during the Cretaceous and most of the coal beds formed during this period are from accumulations of them as opposed to the tree-fern type that predominated in the earlier coal-forming times. The abundant marine life of more primitive type was responsible for the development of oil and natural gas in marine sediments, and Mesozoic beds are responsible for over half the world's production and reserves.

As the Tertiary progressed from 65 m.y. the family of grasses took over large areas of land surface and encouraged the increase and diversification of many of the herbivorous mammals like horses, camels, antelope and deer. Their increase, in turn, resulted in the evolution of new carnivorous mammalian families. Birds probably developed equally rapidly but are much less easily preserved as fossils. In the seas, fishes having the type of bone and fin structure with which we are today familiar outnumbered the older shark class and the clam and snail class became the dominant shelled invertebrates.

Within the class of mammals, the primates are recognised as a distinct family as far back as the late Cretaceous and had diversified into many types by 60 m.y. The earliest evidence of primates using artifacts of bone or stone is found in South Africa and in East Africa in the Great Rift Valley between Latitudes 12°N and 4°S. In these sites, artifacts have been found in association with skulls and other hominid bones. The bones and artifacts can be dated with confidence between 1.5 and 3 m.y. and may go back as far as 5 m.y. before the present.

No human-like remains as old as that have been found so far in any other part of the world. The next oldest evidence is "Java Man", a skull found in about 1890 and dating between 700,000 and 600,000 years before the present. Skulls and parts of skeletons that have been named Sinanthropus or "China Man" which were found in 1921, were dated at about 500,000 years on the basis of tests when the techniques of dating were not too reliable and re-study has been impossible due to the loss of the bones in the confused period of civil war following World War II.

Man was well-established in Europe and Asia by the time the Pleistocene Ice Age had reached its maximum and lived and hunted up to the edge of the northern ice caps which waxed and waned with alternating periods of falling and rising temperatures. There is no evidence that Man, or his ancestors, were in North America before the Ice Age. Recent discoveries have been made of animal bones that have been chipped and fashioned as implements in an area in the northwest part of Yukon Territory, Canada, and extending into the adjoining part of Alaska.

This roughly circular area evidently escaped glaciation when ice caps covered most of the surrounding territory and a complete succession of sediments of the mid-glacial period has been preserved. The bone artifacts found were not in their original position of deposition but had been carried by floods or streams and deposited in beds of sand or gravel which have been dated with confidence, by radioactive tests on volcanic ash layers, at between 30,000 and 35,000 years before the present. Although no human bones have yet been found in the area, these artifacts may place the migration of Man, from northern Asia across the Bering straits and along the Aleutian Islands to the North American mainland, at this time or earlier. The discovery of 30,000 year old human settlements or camping spots is hoped for and expected in this far-north area that escaped the ice caps.

The Influence of Life on Mineral Deposits

The evolution and expansion of life on earth has had an obvious influence on the composition and relative abundance of sedimentary formations especially limestones and other carbonates, and bituminous or graphitic shales. Equally obvious is the dependence on organic matter of all fossil fuels and, to a degree, of phosphates. Less obvious, but supported by increasing evidence, is the significant part played by living organisms, especially bacteria and algae, on the precipitation of metals from solutions. The process may take place from surface solutions at atmospheric temperature and pressure or from hot solutions of igneous origin. Bacterial precipitation may occur on the bottom of lakes or inland seas, for example the currently active precipitation of metals on the floor of parts of the Red Sea. The close association of life with metalliferous solutions on the floors of oceans was spectacularly shown in an illustrated article in the *National Geographic Magazine* (Vol. 156, No. 5) of November 1979. This shows, in two areas of the Pacific Ocean, the abundant visible plant and animal life surrounding vents of mineral-bearing solutions at high temperatures (up to 350°C) and at depths of over 3,000 metres.

Even the precipitation of metals from hot solutions below rock surface in channels or pore spaces may involve bacterial action.

While the extent of the action of primitive organisms on metal deposition is still uncertain, there is no doubt that life is a factor involving both the primary deposition of minerals and the secondary concentration of metals from low-grade sources into proportions that are of value in human economics.

DISTRIBUTION OF MINERAL RESOURCES

One of the characteristics of mineral deposits of economic value, whether solid, liquid or gas, is their irregularity of occurrence over the face of the globe. No country is completely self-sufficient in all mineral supplies—even USSR—and some nations, particularly in western Europe, are almost completely dependent upon imports of raw materials for their energy supply, metal fabrication and agricultural fertilisers. It is sobering to realise that over 50% of the known reserves of oil lie within a few hundred kilometres of the shores of the Persian (or Arabian) Gulf; that about 70% of the known reserves of platinum metals, mineable at current prices, are in a quite limited area of South Africa; that over 50% of the current world tin production is from an area, centred on the Malaysian Peninsula, that is about half of 1% of the earth's surface.

It is, therefore, of strategic and economic importance for anyone even indirectly concerned with mineral supplies, as well as those directly responsible for their discovery and production, to know something of the known concentrations of these resources, the probable reasons for their location and the potential areas for providing future supplies. It is one of the objects of this Atlas to give some idea of the distribution of each important mineral and the geological reasons behind its occurrences where these are known. This is done, in the first place, by marking on the map sheets, in red print, occurrences of individual metals or minerals (including oil and gas) using the Latin symbols for metals (see Table 1) and full names for non-metals. Obviously, on maps of such small scale, only larger *areas* of metal or mineral occurrence can be shown although in a few exceptional cases individual mines or deposits are marked.

Table 1

ATOMIC SYMBOLS OF METALS USED ON MAP SHEETS

Antimony	Sb	Manganese	Mn	Tantalum	Ta
Chromium	Cr	Molybdenum	Mo	Titanium	Ti
Copper	Cu	Nickel	Ni	Tungsten	W
Gold	Au	Platinum	Pt	Uranium	U
Iron	Fe	Silver	Ag	Zinc	Zn
Lead	Pb	Tin	Sn		

It is not possible to indicate on these small maps the size of production or reserves in an area but some idea of the relative importance of a nation or an area in the production of a particular mineral may be obtained from the chapter (p92) where, under each mineral or metal of importance, the leading producing countries are listed with their 1978 production (and 1979 where available) and the estimated reserves for 1979. At the head of each list in this chapter is a brief note on the types and ages of rocks that carry the major proportion of world supplies of the particular resource.

As a background to the notes on individual mineral resources there is a need for a summary, necessarily brief, of the forms of orebodies and the present concepts of their genesis. In regard to forms or shapes of orebodies, there is a popular impression that veins form the object of most metal mining and, indeed, prior to the present century they did supply a large proportion of ores mined underground, especially of gold and silver. Today, veins form only a small proportion of producing orebodies, partly because most of the earlier-found, high-grade veins have been exhausted but more because the increasing use of mechanical methods of mining favours orebodies of larger dimensions, and correspondingly lower grade, that can be worked in openpits or by large underground operations. In line with this trend, a large proportion of metallic deposits in production today fall into two physically and geologically different groups.

The first may be termed "strata-bound" which means that the ore is confined between the upper and lower limits of a particular sedimentary or volcanic stratum. The ore may be actually bedded, such as most iron formations, the copper of the Zambia and Zaire "copper belt", or the "Kupferschiefer" of Germany and Poland. Alternatively the ore may be in irregular masses, sometimes roughly stratified, within a particular bed, as is the case in many of the world's lead and zinc deposits.

The other group of large-scale metallic deposits, in the sense of geometric form, is that in which metals (especially copper and molybdenum) are distributed in small disseminations or fractures within intrusive masses, often roughly circular in plan, of broadly granitic composition. Such deposits, known as "porphyries" (although the containing rocks would often not strictly be classified under this term) have a distribution related to plate boundaries.

Changes in concepts of the origin of many ore deposits have broadly coincided with the general trend towards the mining of larger orebodies with a correspondingly lower grade. Prior to the 1950's, it used to be thought that a great many metallic deposits were formed by hot solutions coming from the interior of the Earth which emplaced the metallic minerals in rocks that had consolidated in a much earlier geologic time. While the metal-bearing solutions are still accepted as initially originating from the deeper layers of the Earth's crust, it is now believed that a large proportion of deposits were formed at or close to the Earth's surface then prevailing; the concentration depending on the physical and chemical conditions of the current environment. Some orebodies, for example, are formed by the precipitation of metals from hot springs, related to volcanic activity, reaching the surface under the water of seas or lakes and encountering temperatures, or chemical conditions, that brought them to the point of saturation of a particular metal. This process of metal precipitation is occurring now in parts of the world, notably in the Red Sea (see North Africa sheet p56). Many of the important deposits of copper, zinc and lead (often accompanied by silver and gold) of the world appear to have been formed at an under-water surface associated with volcanic activity.

In the other cases, metal-bearing solutions seep through permeable sedimentary rocks relatively close to surface and precipitate the metallic minerals in pore-spaces or fractures or by replacement of non-metallic minerals. This process seems to have been responsible for lead-zinc deposits of the Mississippi Valley type.

The study of plate tectonics has shown a relationship of some ore deposits to plate boundaries. The example mentioned in the Red Sea where metals are being deposited today appears to be related to a diverging boundary and many deposits now being mined probably have a similar original relationship. But the most definite and important association of ore deposits with plate boundaries is that of the "porphyry" deposits of copper and molybdenum with subduction zones along converging boundaries. The most important examples of this association are along the Cordilleran mountains of North and South America from which belt porphyry deposits supplied 40% of the 1979 world production of copper and 88% of the molybdenum, but similar examples are associated with subduction zones in the southwest Pacific and in a belt from Iran to Pakistan.

The major proportion of uranium deposits over the world owe their concentration to surface solutions that have leached uranium (a very soluble metal) from rocks with an above-average content. These solutions have seeped through porous beds or open structures and deposited the uranium under favourable chemical or structural conditions. A limited number of deposits may owe their origin to solutions of deeper, magmatic sources. The various classes of uranium deposit are summarised under the section devoted to this mineral in the chapter on Mineral Resources (Production and Reserves) (p92).

Although not yet contributing to world metal production, "manganese nodules" on the floors of oceans may be a future source of some metals. These nodules, commonly between 3 and 15 centimetres in diameter consist of manganese and iron oxides with smaller but significant proportions of nickel, copper and cobalt. They lie loose on the ocean floor in certain areas, the selection of which is not clearly understood. A belt across the Pacific has a higher proportion of nodule concentrations than any other known ocean floor and experiments in recovering the nodules from the considerable depths and treating them for metal extraction are in progress.

Petroleum products, oil and gas, are believed to owe their origin to micro-fauna and -flora that are buried in the accumulation of muds or clays which become shales when compacted by overlying younger beds. As the pressure of the overlying weight of the new sedimentary layers becomes greater, with the coincident increase in temperature, the trapped organic matter begins to disintegrate, beginning when a covering of about 1,200 metres has been achieved. The products of this disintegration process consist of solid organics and oil and gas. Ensuing heat and pressure may squeeze the oil, gas and water enclosed in these shale beds into adjoining sediments that have greater porosity, e.g. sandstones or porous limestones.

The conditions for maximum generation of petroleum include abundant oceanic life on a broad scale, relatively high ocean temperature and a rapidly sinking basin or graben structure permitting the thick accumulation of sediments that "wedge out" at the sides. Because of the temperature factor, a large proportion of major oil fields even though they may be in relatively high latitudes now, were within 30° of the Equator at the time of the accumulation of the oil.

In relation to geological time, the following are breakdowns of known petroleum reserves in 1971 into geological eras since Precambrian:-

(1)	pre-220 m.y.	117,000	million barrels	11.4%
(2)	between 220 and 65 m.y.	670,000	,, ,,	65.4%
(3)	post-65 m.y.	238,000	,, ,,	23.2%
		1,025,000	,, ,,	100.0%

However, the above figures are heavily weighted by the reserves of the Middle East which constitute over 50% of those of the whole world and nearly all these reserves were formed between 110 and 90 m.y. If we exclude this unusual concentration, the time distribution of the remaining world reserves is much more even, i.e. pre-220 m.y.—27%; 220-65 m.y.—39%; and post-65 m.y.—34%. The pre-220 m.y. reserves have probably suffered more than those of later periods by their destruction in the course of the subduction of some continental shelf areas and the dissipation of oil and gas during mountain-building activity.

In the above figures, oil sands (or "tar sands") and oil shales are not included. These are of immense potential value, increasingly so as the price of oil from conventional sources rises. In both oil sands and oil shales, it is the recovery that is the major engineering problem. This is indicated by such figures as those of western Canada where the oil indicated by drilling the Athabasca and related oil sands exceeds 700 billion barrels, or 93 billion tonnes, (more than the total world conventional oil reserves in 1978) but only 22 billion barrels are likely to be recoverable by mining. Oil shales, containing less oil per cubic metre, present still greater problems in mining, recovery and disposal of the immense quantities of waste but are under serious consideration.

Coal and lignite beds of all grades are formed originally from the accumulation, burial, compaction and alteration of land plants. The oldest coal beds, about 320 million years old, were formed from the tree-fern type of plant whereas the younger coals, formed from deposits dating 180 million years old or later, are accumulations of deciduous plants for the most part. The productive coal beds of western Europe and eastern North America are of the older tree-fern type while those of western North America are of the younger deciduous type. In Asia and Australia there is not the same break in coal generation.

INTRODUCTORY COMMENTS
TO THE MAP SHEETS

The following, and principal, part of the Atlas consists of seven double and three single page map sheets that collectively cover all the land areas of the world with the exception of some oceanic islands. In the selection of the area to be covered by each sheet, the main consideration was to cover the desired area without breaking natural geological features such as mountain ranges. For the same reason, the long axis of a sheet is sometimes orientated diagonally to latitudes e.g. the Northern Europe sheet is orientated in a northeasterly direction to include most of the Caledonian folding of Britain and Scandinavia; the Indonesia-Australasia sheet in a southeasterly direction to include the Indonesian trenches and folding.

Limitations imposed by page-size and economics have made it impracticable to keep the map sheets on a constant scale and those used vary between limits of 1:7.5 million and 1:30 million. Similarly, no one type of projection is equally suitable for all parts of the world. Projections that distort and exaggerate land shapes in polar regions, such as the Mercator, have been avoided and those that give the truest outlines for each part of the world have been chosen—in most cases forms of conic projections. In the case of the sheet covering North and South America the convention of placing north at the top of the sheet has not been followed.

Each map sheet is accompanied by notes that summarise the geological history of the respective part of the world. Geological events, whether sedimentation, orogenies, plate movements, life or significant mineral concentrations, are discussed within a time frame. In addition some notes are placed on the maps themselves to draw attention to special geological events or features and, in some cases, to the effects on human history.

As far as possible ages of rocks in millions of years (m.y.), obtained from radiometric determinations, have been used on the maps in preference to formational or time names. Ages without brackets indicate the date of the origin of the indicated formation; those bracketed indicate the date of metamorphism (usually accompanying an orogeny) that restarted the radioactive clock.

On the map sheets, *areas* of mineral deposits are shown in red print by the respective atomic symbols in the case of metals and by full names of non-metals. It is obviously impossible on maps of such small scale to mark by name individual deposits, even important ones, where these occur in clusters like the Witwatersrand, Porcupine and Kalgoorlie gold areas, the Arizona copper area and any coal basin. On the other hand, mines of importance situated by themselves, like Homestake, Broken Hill NSW, Tsumeb, Bingham or Kidd Creek can be, and are, marked by name.

It has not been found feasible to be entirely consistent in marking on the maps oil, gas and coal fields. For example, on the Northern Europe sheet, individual oil and gas fields in the North Sea have been indicated, with some generalisations, and the same procedure is followed in the producing areas in and near the Arabian Gulf (African and Southern Asia sheets). In North America and USSR, on the other hand, where smaller scales are used and where there is more overlapping detail, individual oil fields are not emphasised; rather areas that contain numerous oil or gas fields are shown within a red boundary.

On maps of these scales it has not been practicable generally to give any indication of the relative importance, in output or reserves, of individual deposits or areas of mineralisation. In the Mineral Resources chapter (p92), a list is presented for each major mineral resource showing, in approximate order, the countries that played a major part in recent production and, where available, some indication of reserves. By comparing these lists with the maps some idea may be obtained of the relative importance of mineral areas marked within the boundaries of a particular country.

National boundaries have been drawn in order to aid in locating geological features but they are not claimed to be accurate in detail nor should they be taken to have any political significance..

MAP	SCALE	PROJECTION	PAGES
The Americas	1:27,300,000	Bipolar Oblique Conic	34-35
Northern Europe	1: 7,500,000	Lambert Conformal Conic	46-47
Mediterranean Europe	1: 7,750,000	Lambert Conformal Conic	50-51
Northern Africa	1:17,000,000	Chamberlain Trimetric	56-57
Southern Africa	1:17,000,000	Chamberlain Trimetric	61
USSR	1:19,360,000	Conical Orthomorphic (Modified)	66-67
Southern Asia	1:20,000,000	Lambert Conformal Conic	70-71
Indonesia-Australasia	1:21,000,000	Lambert Modified Equal Area	82-83
Antarctica	1:30,000,000	Polar	88
Arctic	1:30,000,000	Polar	91

THE AMERICAS SHEET

The rather unconventional aspect of the North and South American continents, and the fact that the scale of the map is smaller than any of the others in this Atlas, is necessary if we are to avoid carving up the continents and thereby breaking natural geological units. One of these, the Cordilleran Range, extends from Cape Horn to the Aleutians, a distance of over 20,000 km, and is the longest continuous continental structure. The Cordilleran Range is the connecting link between the North and South American continents which, in other ways, have rather different geological compositions although both are built on Precambrian foundations around which later sedimentary and volcanic formations have been laid.

North and South America are separated geologically by a small tectonic plate which includes the Caribbean Islands and part of Central America and has been named the Caribbean Plate. This involves a later episode in geological history compared to that of the North and South American continents and so it seems logical to summarise separately the geological history of North, South and Central America (including the Caribbean Islands).

NORTH AMERICA

A glance at the map shows that the North American continent and its adjacent Arctic islands, together with Greenland (which was part of the continent until comparatively recent geological times) is built around a foundation of Precambrian rocks with the oldest (Archaean) occupying a roughly central position surrounded by younger folded and unfolded formations. The latest Precambrian folding on the southeast side of the Shield, and the still more extensive Appalachian folding that is shown on the map to extend from east and north Greenland to where it is covered by later sediments in Alabama, testifies to a long history of alternate convergence, collision and parting with the Eurasian and African continents. The west side of the continent is bounded by the Cordilleran Range which resulted from the westward convergence of the plate carrying the American continents onto the Pacific plate.

The chronological history of geological events, and the formation of major mineral concentrations, are summarised below.

Pre-2,500 m.y.

These early Precambrian rocks form one major craton, or platform, (commonly known as the "Superior Province") that occupies the central part of the Canadian Shield from Longitude 68°W to Longitude 96°W where it becomes hidden by much younger sediments. A second, much smaller area of approximately the same age, pre-2,500 m.y., occurs north of Great Slave Lake. A third, of which only remnants are definable, is on the west side of Greenland and on the part of Labrador to which it was formerly attached. This area includes the oldest dated rocks in the world (3,790 m.y.). A fourth Archaean area, sometimes known as the "Wyoming Province", extends from that state into Montana and South Dakota but is exposed only in limited "windows" through younger cover rocks.

The history of these ancient cratons is more speculative than that of later stages of the Earth's history but it appears that volcanic activity predominated while sedimentary formations consisted mainly of reworked volcanic material. Modern type sediments, resulting from the weathering and breaking down of older rocks, were not formed at this time, possibly because the Earth's atmosphere still lacked appreciable proportions of oxygen. In spite of this, relics of life of simple blue-green algae types have been found in North America in rocks that date back to 2,700 m.y.

On the map it may be seen that these early Precambrian areas show a "grain"— in the case of the larger craton, easterly; in the one north of Great Slave Lake, northerly— caused by narrow belts of volcanic-sedimentary rocks (coloured purple) separated by broader areas of gneiss and granite (coloured pink). This orientation may result from the volcanic activity being concentrated along troughs that were controlled by the fracture system formed on the original crust, or possibly from ancient tectonic plate movements on a smaller scale than those of later Earth history. Whichever was the predominant control, the folding that culminated in the great orogeny at about 2,650 m.y., accompanied by granite intrusion, seems to have followed these trends. Orogenies of approximately the same period have occurred in most of the larger Precambrian areas of the world. Erosion since that time has exposed the roots of the folded and faulted layers, and the granite intrusion and granitisation involved in the orogeny, leaving the volcanic-sedimentary belts (commonly known as "greenstone belts") separated by wider layers of gneisses and granites. The latter may include reworked parts of the original continental crust.

Mineralisation of economic importance was associated with the early Precambrian volcanic activity. This includes the **gold** deposits that at one time formed the basis of Canadian mining activity as well as **copper-zinc** (with associated gold and silver) sulphide bodies that were probably deposited around volcanic centres by sub-marine hot spring activity between successive volcanic flows. Examples of these are in the Noranda area of Quebec and at Kidd Creek, Ontario, the most outstanding of them all. Also deposited between periods of lava outpouring were **iron** formations, relatively low-grade but susceptible to concentration as commercial operations. **Nickel** mineralisation occurs in Canada in basic volcanic or intrusive rocks but is relatively less important than in rocks of this age in Australia and Zimbabwe. The important nickel deposits of Canada are in rocks of the following period.

2,500 to 1,800 m.y.

The extensive major orogeny that ended the first recorded stage of geological history in North America resulted in the first stable cratons, or platforms, which must have carried ranges of high and rugged mountains that were subject to rapid erosion. It has been suggested by some specialists in Precambrian geology that there was a relatively rapid change about this time to an atmosphere containing about the same proportion of oxygen as it does today. Although such an event would explain certain geological occurrences there is no positive proof of such drastic change. Certainly, however, it is after this time that the weathering of existing rocks began to yield sediments of the cleanly-separated type, e.g. sandstones, clays or shales and limestones, that we see in subsequent geological successions up to the present.

The earliest sediments of this period that have been dated are about 2,200 m.y. in North America (although equivalent sediments in Africa are older) and occur flanking the cratons of earliest Precambrian which had by then undergone considerable erosion. Examples of such sediments that have escaped the effects of later orogenies are seen on the north shore of Lake Huron, Ontario, southwest of Lake Superior and on the east side of the Canadian Shield. Originally these sediments, with interbedded volcanic layers, probably ex-

tended over a large part of the present continent (including Greenland). Life continued to evolve and the relative rarity of fossils is due to the lack of hard parts that could be easily preserved rather than to any paucity of marine life. An ice age occurred about 2,160 m.y. as seen from fossil glacial till in the vicinity of Cobalt, Ontario but it probably extended over a much wider area.

Several classes of mineral deposits of major significance were formed during this period of sedimentation and vulcanicity. The **nickel** deposits of the Thompson Belt, Manitoba occur in or close to ultra-mafic intrusives in post-2,500 m.y. volcanics and sediments close to their boundary with the older craton to the south. Other base metal deposits, associated with the volcanic activity, occur in the area extending west and north from Thompson and also to the east in northern Quebec. One of the most important features of economic interest in this period is the deposition of **iron** formations over a great linear distance in eastern Quebec and Labrador and in Minnesota and Michigan where they were the foundation of the steel industry of USA. Since similar iron formations, dating about the same period, are found in many other parts of the world it has been suggested that the change of composition of the atmosphere to an oxygenated one may have been responsible for the precipitation from the seas of iron that the water could no longer retain under oxidising conditions. A third important metallic occurrence of this period is seen in the **uranium** deposits of Elliot Lake, Ontario which were deposited, probably from surface waters, in near-shore conglomerates which date about 2,200 m.y. The veins of **silver** at Cobalt, Ontario, at one time the largest producing area in the world, were formed about 2,000 m.y.

The greatest known concentration of **nickel** (accompanied by **copper** and **platinum** metals) is at Sudbury, Ontario and is associated with a bowl-and-stem-shaped intrusive of mantle origin, dating about 2,000 m.y., which is widely believed to have been triggered by the impact of a large meteorite. In the west, the **gold** ores of the famous Homestake Mine, South Dakota, are in sediments dating about 1,800 m.y.

This phase of North American history ended with an orogeny that affected a vast area surrounding the early Precambrian cratons. The radioactive dating of the major alteration and intrusion that accompanied the folding and faulting of this orogeny varies somewhat from one part of the continent to another but averages about 1,700 m.y. On the map this dating (in brackets to show that it is one of an orogenic episode) may be seen extending from the northwest part of the Canadian Shield, easterly to Baffin Island and southerly to Labrador and eastern Quebec and the corresponding parts of Greenland. Evidence of its extension to the south is limited by the effects of later orogenies and by the covering of later formations but it is probable that the orogeny originally affected a large proportion of the continent.

1,700 to 850 m.y.

Once again the continent was left with ranges of mountains on a still more widespread scale than the previous episode. The vertical elevation, relative to sea level, that always accompanies the later stage of a period of folding, induced rapid erosion and corresponding deposition of sediments. The sediments, and interlayered volcanics, were probably very extensive, being found erratically from the east coast of the continent westerly to the present east flank of the Cordilleran Range. Exposures of these relatively unaltered sediments are seen in several parts of the Canadian Shield, e.g. on the north and south shores of Lake Superior, in the Athabasca Basin of Saskatchewan and in the Northwest Territories. In western USA and Canada, the Beltian Series of sediments was probably deposited between 1,600 and 800 m.y. but has been involved in successive orogenies from late Precambrian until the latest Cordilleran disturbances.

An important aspect of the contact between the lowest sediments of this period and the older eroded basement is the local concentration of **uranium**. Deposits of commercial tonnage and grade, some very high grade, are found in northern Saskatchewan on the edges of, and below the 1,400 m.y. sandstones of the Athabasca Basin. Similar conditions, but less well explored, occur in the Baker Lake area of the Northwest Territories. While most of the deposits lie close to the unconformity of relatively unaltered sediments on older basement, the concentration to the present high-grade deposits appears to have occurred in several stages, some of which pre-dated and others post-dated the unconformity.

In the west of the continent, the outstanding ore deposit of this age is the Sullivan **lead-zinc-silver** mine at Trail, British Columbia, which is in a sedimentary formation, interbedded with volcanics, dating 1,340 m.y. Extensive zinc-lead mineralisation has recently been found in carbonate sediments of 1,000 to 700 m.y. in northwest Canada.

The former famous **zinc** mine of Franklin Furnace, New Jersey, giving radioactive dates of about 950 m.y., and some other **zinc-lead-copper** deposits in eastern USA and Canada lie in sedimentary or volcanic formations of pre-1,200 m.y. that were involved in the Grenville orogeny.

In the Lake Superior region, about 1,100 m.y., a period of extensive volcanic activity occurred with which are associated the **copper** deposits (in many cases in the form of native copper) which were the object of the first major mining activity on the North American continent and which, for a period, dominated the world supply of the metal.

In the northern part of the continent, the sediments and volcanics just described have remained unaffected by any major orogeny up to the present day. In the southern and southeastern half of the continent, however, there was much more disturbance in Precambrian times after 1,800 m.y. A period of intrusion and metamorphism, termed the "Elsonian Event", affected the area south of the present Great Lakes in the period between 1,500 and 1,300 m.y. After this, commencing about 1,200 m.y. and continuing to 850 m.y. a major orogeny known as the Grenville affected the east side of the continent producing a composite fold belt orientated in a northeasterly direction. The folding and alteration involved the sediments and volcanics deposited on the flanks of the 1,800 m.y. craton and also the basements (of several ages) on which they lay.

The Grenville tectonic province, 400 km wide, can be traced from the mid-continent, southwest of which it is hidden by younger formations, northeasterly for 1800 km to the Labrador coast and on to southern Greenland. Rocks of similar age and character are found in Northern Ireland, western Scotland and Scandinavia. The folding, faulting, intrusion and metamorphism, which are believed to have been caused by the collision of the European continent with the North American, continued in surges until about 850 m.y. A characteristic of the Grenville province is the abundance of large rounded intrusive masses of anorthosite, a mafic rock consisting dominantly of calc-alkaline feldspar. An important deposit of ilmenite, ore of **titanium**, is associated with one of these intrusions on the north side of the St. Lawrence River. The magnetic **iron** ore mined from within the Grenville province near its northwest "front" close to the Quebec-Labrador boundary is older than the orogeny and is the altered extension of the same iron belt that is traced from Labrador to the northern part of the Ungava Peninsula.

850 to 570 m.y.

The Grenville orogeny produced what must have been an immense mountain range bordering the southeastern edge of the Precambrian continent. Even today, although much eroded, the Grenville area shows much more difference in elevation than the older parts of the Shield. Sediments derived from the Grenville mountain chain were deposited during this period and remnants are found in various areas near the edge of the Shield. Some of these have been involved in the subsequent Appalachian foldings.

In the west, Precambrian sediments are found to have been folded in an orogeny dated 730 m.y. but its effects, from Mexico to Alaska, have been overprinted by the much later Cordilleran folding.

570 to 225 m.y.

After the end of the Precambrian there was a long period of flooding of much of the continent by seas (continental seas, not oceans) with resulting deposition of great thicknesses of sediments. These consist largely of shales and limestones which contain an abundance of fossils since a surge of invertebrate life of more advanced forms with shells or other hard parts, took place at the start of this period. The large areas of Palaeozoic formations were formed at this time.

Four important classes of mineral deposits were formed during this period:-

(1) **Lead** and **zinc** deposits that constitute a large proportion of the continent's reserves of these metals were formed in marine sediments that are dated between 550 and 290 m.y. By far the greater part of the United States' reserves of these metals are in carbonate sediments of two periods—530 to 470 m.y. and 350 to 290 m.y.—in the main Mississippi basin. Lead and zinc deposits also occur in sediments of the same periods that have been involved in the Cordilleran folding from Mexico to Alaska. Possibly the most important of these deposits, still not in production (1980), is that of Howard's Pass in sediments of about 430 m.y. in the Yukon Territory, Canada. In the Arctic islands of Canada, lead and zinc deposits have been found in relatively unfolded sediments of about the same age, only one of these being prepared for production.

(2) **Coal** began to be formed about 330 m.y. in the east part of the continent following the withdrawal of the seas and the expansion of land vegetation. Coal of this age forms the bulk of reserves in eastern USA and Canada (but less than 35% of total reserves on the continent).

(3) The **potash** deposits of Saskatchewan, which constitute the largest known world reserves, were deposited in cut-off arms of the sea in the period 380 to 350 m.y.

(4) **Oil** and **gas** in sediments of this period have provided a substantial proportion of production to date but a much smaller proportion of remaining reserves on the continent. The oil fields in the east-central part of the continent and many of those on the east flanks of the Cordillera in USA and Canada are in pre-225 m.y. old sediments. For example, 29% of Canada's reserves are in Palaeozoic sediments.

In the eastern part of the continent a tectonic disturbance, extensive geographically and in time, took place with a northeasterly trend and formed what is broadly termed "Appalachian folding". The trend of this folding, which occurred in several pulses, is sub-parallel to the last orogeny of the Precambrian and there is even a suggestion that there may have been some overlap between the dying effects of the Grenville folding and the incipient movements of the new disturbance. Whereas there is some doubt as to the origin of the Grenville orogeny, there is general agreement that the cause of the Appalachian orogenies was the collision of the African and Eurasian continents with the eastern edge of the North American continent (using the term to include the continental shelf beyond the present land surface). Greenland had still not split off from the North American continent and the European continent (at that time possibly separated from the Asian) appears to have moved to such a position that Scandinavia and Britain, which had then not parted, were brought up against the present eastern side of Greenland and Spain-Portugal against the continental shelf east of Newfoundland. At the same time, north-western Africa was probably brought up against the continental shelf off the eastern shore of Nova Scotia, New Brunswick and the eastern United States.

The folding that resulted from this collision, which affected an area from south-west of Alabama to Newfoundland and on from there along the east and north coasts of Greenland, appears to have taken place in several phases between 500 and 320 m.y. with some temporary reversals to the convergence. The parting, in such reversals, usually occurs along a line different from that of the preceding collision. These reversals, or periods of rebound, permitted sediment deposition to occur between pulses of folding, intrusion and alteration. In some places along collision "sutures" parts of the ocean floor were thrust up on the eastern edge of the continent and such remnants are now recognised as "ophiolites", particularly in Newfoundland.

A straight lineament, roughly along the southwestern edge of the folding and extending 1,600 km from just west of Albany, New York, to Alabama, is indicated by magnetic and gravity evidence. Movement on this apparent fault seems to have been mainly lateral and may be comparable to that on the Rocky Mountain-Tintina fault in the Canadian Cordillera or to the Altyn Tagh fault in China.

The folding was accompanied by granitic intrusion which was concentrated in three phases dated about 500 m.y., 400 m.y. and 320 m.y., respectively.

The waning stages of the Appalachian folding and intrusion are dated at about 320 m.y. and were followed by subsidence and marine incursions into the centre of the continent resulting in sedimentation which then continued for the next 100 m.y.

In the area of the present Cordilleran Range, marine sediments had been deposited on the continental shelf from 570 to 400 m.y. with minor tectonic disturbances. In Devonian times (395 to 345 m.y.) there was evidently a period of major block faulting but no significant folding appears to have taken place in the west between that of late Precambrian (about 720 m.y.) until approximately 225 m.y.

225 to 65 m.y.

Sedimentation, and some volcanic activity, continued in the central and southern parts of the continent and it seems probable that the belt of Appalachian folding was, to a large extent, above sea level for most of this period. In the west, marine sediments from invasions of the sea alternated with freshwater sedimentation from large inland lakes. It was on the shores of these that land vertebrates, dominated by the dinosaur group, developed on this continent and large numbers of their fossil remains are found in western Canada and USA.

On the eastern margin of the continent, the important event of this period was the rifting and drifting apart of the North American continent from the Eurasian and African continents to which it had been attached at the respective continental shelves. The initial zone of rifting, that subsequently became the Mid-Atlantic Ridge, started between South America and Africa at about 180 m.y. The tearing apart, rather like someone tearing in two a piece of paper lying flat on a desk, started in the south and extended north producing a wedge-shaped ocean with the separation of Greenland and Britain/Ireland from Canada not occurring until about 80 m.y. The parting of Greenland from Eurasia took place still later, about 60 m.y.

The westerly drift of the plate carrying the North American continent resulted in it being underthrust, in the process known as subduction, by the thinner oceanic crust of the Pacific Plate. One of the first effects of a subduction event is the production of a trench, or system of trenches, off the continental shelf and parallel to it. This trench sinks, allowing an accumulation of sediments usually accompanied by volcanic material. As the convergence of the two plates proceeds with the oceanic plate plunging beneath the continental, the accumulated sediments and lavas, being less dense than the oceanic crust, tend to be subducted less and to be crumpled and shortened in the direction of convergence. In the Cordillera the first stage of the orogeny as a whole started over 225 m.y. ago, as dated by granites, but the main shortening of crust began about 100 m.y. ago, and continued, in pulses, for the next 30 to 40 million years. The shortening took the form of folding and over-folding accompanied by thrusting on a succession of planes.

Accompanying the folding, faulting and metamorphism in each pulse was the extensive outpouring of lavas at surface and granite intrusion (and granitisation of sediments) at depth. From the map it can be seen that granitic rocks, exposed by erosion, occupy a large proportion of the more westerly units of the fold belt and that volcanics of this period (green with "V" pattern) cover extensive areas in the belt east of the granite one.

North of the USA-Canada border, the Cordilleran fold belt on the mainland is fairly distinctly divided into four subsidiary north-west-trending belts which decrease in age from west to east. These are:-

(1) The Coast Range, mainly granite, of about 205-140 m.y. age;

(2) The Intermontane belt, probably brought up from the south by horizontal movement, of about 160 m.y. age;

(3) The Omenica Crystalline belt, of about 145 m.y.; and,

(4) The Rocky Mountain belt, nearly all older sediments but the folding dates from about 90 to 55 m.y.

South of the international border these divisions are not as clear due to a large proportion of the area being covered by later lava flows and sediments. In general, however, the same trend prevails from older folding, faulting and intrusion in the west to younger in the east.

A major tectonic feature of the Cordillera is horizontal movement, on a tremendous scale, roughly parallel to the fold axes. The most extensive structural unit recognised in the Cordillera is the fault, or fault system, of the Tintina-Rocky Mountain Trench which is known to extend from a point 200 km west of the Alaska-Yukon border southeasterly across the Yukon Territory and down through British Columbia to south of the USA-Canada boundary, a total distance of 2,400 km. The horizontal movement on this great structure is "right-handed", i.e. the southwest side has moved northwesterly relative to the opposing side. In the northern, Tintina, part of the structure, this sliding movement is now believed to have amounted to not less than 450 km and possibly over 500 km. South of Latitude 55°N the movement seems to be taken up to a large extent on a branch that runs in a more southerly direction and is known as the Columbia River Fault although about 80 km of right-handed movement still appears to have taken place on the Rocky Mountain Trench fault itself. This horizontal movement, and many other examples of lesser magnitude, occurred in mid-Cretaceous time, about 100 m.y. but similar movement continued in the Tertiary.

The most important mineral deposits formed during this time interval in North America are coal, copper and uranium. **Coal** beds were formed in the later stages of this period, extending into Tertiary time, on the margins of fresh water bodies in a belt just east of the Cordilleran folding but affected by minor folding and faulting in its later stages. The period 110 to 20 m.y. includes the formation of 65% of the coal and lignite reserves of USA and a still larger proportion of the more modest Canadian reserves.

The principal source of **copper** today in western North America is in the form of disseminations in "porphyry" intrusions—a generalised term covering several varieties of granitic rocks—or in altered lavas or sediments close to such intrusives. Porphyry copper deposits form a belt extending from South America, through Mexico, western USA and Canada and are related geographically and in origin to the subduction zone at the western margin of the American Plate. The important porphyry coppers of Arizona seem to diverge from this zone, being farther inland than the main belt. These porphyry copper deposits appear to have formed in several rather distinct periods, starting in the time interval under discussion but extending on into the next one. Thus, the copper deposits in the southern part of British Columbia date between 200 and 160 m.y. while those in Arizona range between 80 and 60 m.y. Still younger deposits are referred to in the notes following.

Lead and **zinc** deposits of this period tend to occur in small, but often high grade, bodies in Mexico and western USA The silver-lead veins near Mayo, Yukon, are in sediments of about 120 m.y. and were probably deposited about 75 m.y.

In western USA **uranium** occurs in sediments, mainly of freshwater origin, from about 220 m.y. to the end of this time interval and beyond it. The metal is believed to have been leached by surface waters from granitic outcrops or from high-silica lavas or ash beds and then precipitated in beds carrying organic trash or having contact with hydrogen sulphide leaking up fault zones. The Colorado Plateau, a basin shaped structure covering about 13,000 km^2, supplies about 60% of USA uranium production.

Oil and **gas** were generated in marine sediments on the east side of the Cordilleran belt from Mexico to northern Canada but in many cases have been dissipated by subsequent folding and faulting. The generation of petroleum extended beyond the end of this time interval into the Tertiary, especially in Canada. The importance of this stage is indicated by the fact that 70% of the Canadian oil reserves are in sediments of Mesozoic and Tertiary age. Some of the porous beds of Mesozoic age, however, are the containing formations of oil and gas originating in older formations. An outstanding example is that of the Athabasca Tar Sands at Fort McMurray, Alberta. While the beds there are Cretaceous (about 110 m.y.) the oil has almost certainly migrated into them from older, underlying sediments that were the source rocks. The escape of the volatile products has left the viscous "tar".

65 m.y. to Present

In the western part of the continent, the Cordilleran orogeny, although past its peak, continued with folding, faulting, granitic intrusion and periodic extensive volcanic activity. On the map large areas of volcanic material ranging in age from 50 m.y. to the present may be seen. Some areas of recent volcanic activity are the sites of **geothermal power** that is currently receiving increasing attention.

Tectonically the same massive lateral fault movements noted in the previous time interval continued along the length of the Cordillera. An example is the Denali fault that follows a curved line from the southeast end of the Alaska panhandle, through southern Alaska to its west coast for a distance of at least 2,200 km. The horizontal movement on this fault is "right-handed", like that of the Rocky Mountain Trench, and has resulted in the Pacific side being moved westerly for about 300 km relative to the opposing side. There are many other faults with major horizontal movement in the Cordillera, the best known being the San Andreas, passing through San Francisco, on which movement amounting to 260 km can be shown to have occurred since 25 m.y. and is still continuing.

The thickening of the crust that resulted from the intense folding and thrusting during the most active part of the orogeny—since this crust is "floating" on mantle material—caused an increase in elevation of the whole Cordilleran Range, with consequent rapid erosion. On the eastern side of the mountains the eroded material was deposited partly in freshwater lakes and partly in continental seas that swept in from the Arctic and from the south. This resulted in marine sediments being interbedded with terrestrial sands and clays.

A period of volcanic activity took place mainly, but not entirely, in the most recent 17 million years of geological history. The greatest extent of these volcanic deposits is in a large oval area, known as the Columbia River Plateau, covering most of the State of Oregon, the southern part of Washington and the northern parts of California and Nevada. Leading into this area, from the east, is a belt of 2 million-year and younger volcanics that extends from Yellowstone Park, Wyoming, where there is still geyser activity, and which may be the site of a "hot-spot".

An extensive north-south fault zone, known as the Rio Grande Rift, runs from central Colorado to the Mexican border and has had volcanic activity associated with it in several places and currently hot spring activity.

Other sites of volcanic activity in the past 20 million years in North America are in central British Columbia, Yukon Territory, and Alaska and in western and southern Mexico, the last being still active.

Steam and hot water associated with former volcanic activity is developed for geothermal power in Geyser, north of San Francisco, and in Mexico and is under development in Utah, southern California, Nevada and near the Rio Grande Rift.

Economic deposits of a wide variety were formed during this time interval. Directly associated with the later stages of the Cordilleran orogeny are the younger examples of "porphyry" type **copper** and **molybdenum**. Examples include the historic Bingham Canyon copper mine and the Climax molybdenum mine in Colorado. Of about the same age, and related origin, are the **gold** and **silver**, often associated with base metals, vein-type deposits throughout the length of the Cordillera which first attracted European exploration and later inspired the first settlement of the American west. Those of the Tonopah area, Nevada, are examples. The majority of the **silver** veins in Mexico, in 1979 the largest world producer of the metal, were formed at about 30 m.y.

Some of the **uranium** of western USA is in Tertiary freshwater sediments, especially of the period 50 to 40 m.y. The **phosphate** beds of Florida and the Carolinas result from fossilised faunal material in marine sediments of about 35 to 24 m.y. age. Salt beds, gypsum and sulphur are especially abundant in beds dating from 22 to 8 m.y.

Much of the **oil** and **gas** of the Gulf of Mexico, both on- and off-shore, and of coastal California and Mexico, were formed during this time interval, although part of what is found in beds of 65 m.y. and younger has migrated in from older source-beds.

An economic result of the action of surface waters in recent geological history was the production of **alluvial** or **placer gold** which played an important part in the early development of the western USA. With the exception of one locality, the placers that must have developed in Canada and northern USA were dissipated by the ice sheets that swept south in the Pleistocene Ice Age. The exception is the Klondike area in the Yukon Territory, and the adjoining part of Alaska which, for some local climatic reason, escaped glaciation, thus allowing the preservation of rich river gravels that made this area famous at the end of the 19th century.

The last major geological event was the Pleistocene Ice Age, referred to above, that lasted from between 2.0 and 1.5 m.y. until 10,000 years ago, by which time the ice caps had disappeared from the continent except for a few remnants in Baffin Island and Ellesmere Island and the ice cap that covers the larger part of Greenland. The continental ice, which waxed and waned with the alternating glacial and interglacial phases, at its maximum reached a southerly line passing just south of the Canada-USA border from the Pacific coast to the east side of Montana; then southeasterly to about St. Louis, Missouri, easterly to Cincinnati, Ohio; then on a sinuous northeasterly line to the Pennsylvania-New York State boundary east of Lake Erie and thence to the Atlantic coast just south of New York City.

The thickness of the ice sheet at its maximum was over 3 km and the weight resulted in the depression of the northern and eastern part of the continent relative to sea level. At the same time, however, the "locking up" of a significant proportion of the surface water of the world in ice sheets resulted in the world sea level being lowered to over 100 metres below that of the present day.

SOUTH AMERICA

A glance at the map shows that Precambrian rocks occupy a considerable proportion of the north and east part of the sub-continent, concentrated in three obvious areas, with a fourth partly obscured. Each is a craton or stable nucleus

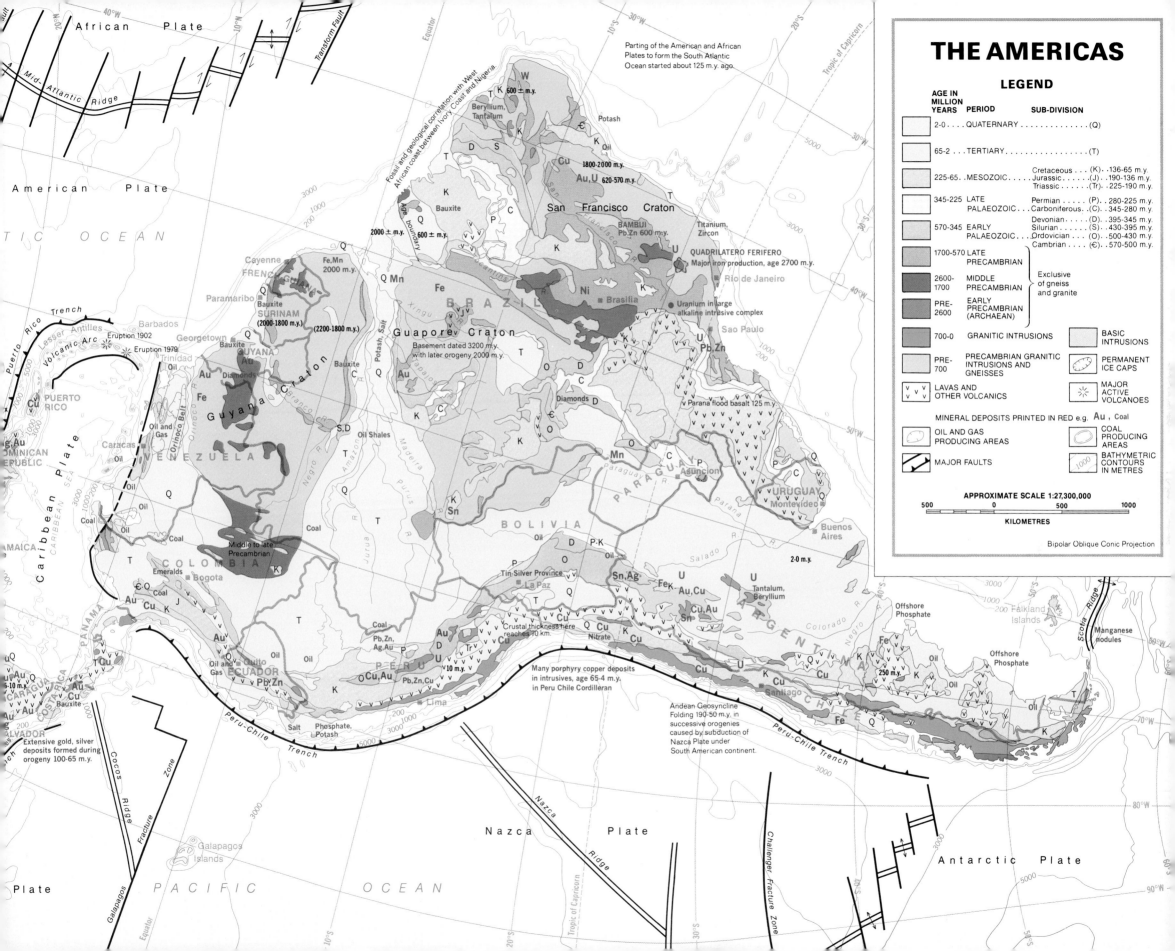

THE AMERICAS

LEGEND

AGE IN MILLION YEARS	PERIOD	SUB-DIVISION
2-0	QUATERNARY	(Q)
65-2	TERTIARY	(T)
225-65	MESOZOIC	Cretaceous (K) 136-65 m.y.
		Jurassic (J) 190-136 m.y.
		Triassic (Tr) 225-190 m.y.
345-225	LATE PALAEOZOIC	Permian (P) 280-225 m.y.
		Carboniferous (C) 345-280 m.y.
570-345	EARLY PALAEOZOIC	Devonian (D) 395-345 m.y.
		Silurian (S) 430-395 m.y.
		Ordovician (O) 500-430 m.y.
		Cambrian (€) 570-500 m.y.
1700-570	LATE PRECAMBRIAN	
2600-1700	MIDDLE PRECAMBRIAN	Exclusive of gneiss and granite
PRE-2600	EARLY PRECAMBRIAN (ARCHAEAN)	

700-0 GRANITIC INTRUSIONS

BASIC INTRUSIONS

PRE-700 PRECAMBRIAN GRANITIC INTRUSIONS AND GNEISSES

PERMANENT ICE CAPS

v v v LAVAS AND OTHER VOLCANICS

MAJOR ACTIVE VOLCANOES

MINERAL DEPOSITS PRINTED IN RED e.g. Au , Coal

OIL AND GAS PRODUCING AREAS

COAL PRODUCING AREAS

MAJOR FAULTS

1000 BATHYMETRIC CONTOURS IN METRES

APPROXIMATE SCALE 1:27,300,000

500   0   500   1000
KILOMETRES

Bipolar Oblique Conic Projection

African Plate

American Plate

Mid-Atlantic Ridge

Transform Fault

Parting of the American and African Plates to form the South Atlantic Ocean started about 125 m.y. ago.

Fossil and geological correlation with West African coast between Ivory Coast and Nigeria.

K 600 ± m.y.

W

Beryllium, Tantalum

Potash

Cu

Oil

Au,U 620-570 m.y.

1800-2000 m.y.

San Francisco Craton

Titanium, Zircon

BAMBUI Pb,Zn 600 m.y.

QUADRILATERO FERIFERO Major iron production, age 2700 m.y.

Rio de Janeiro

Ni

Brasilia

Uranium in large alkaline intrusive complex

B R A Z I L

Sao Paulo

Pb,Zn

600 ± m.y.

2000 ± m.y.

Age boundary

Cayenne FRENCH GUIANA

Fe,Mn 2000 m.y.

Q Mn

Fe

Guapore Craton Basement dated 3200 m.y. with later orogeny 2000 m.y.

Xingu R.

Tapajos R.

Diamonds

Parana flood basalt 125 m.y.

Paramaribo

Bauxite

SURINAM (2000-1800 m.y.)

(2200-1800 m.y.)

Potash, Salt

Au

Georgetown

Bauxite

GUYANA

Oil

Au

Diamonds

Guyana Craton

Fe

Branco R.

Negro R.

Amazon R.

Madeira R.

Purus R.

Jurua R.

S,D

Oil Shales

Trinidad Oil

Mn

C

P

PARAGUAY

Asuncion

URUGUAY Montevideo

Sn

Coal

Oil

BOLIVIA

Oil

Buenos Aires

Salado R.

2-0 m.y.

Caracas

VENEZUELA

Oil and Gas

Orinoco Belt

Oil

Oil

Coal

Coal

Middle to late Precambrian

Emeralds Bogota

COLOMBIA

€O Coal

Au Cu

J

Oil and Gas

Quito ECUADOR

Pb,Zn

Oil

PERU

Coal

Pb,Zn, Ag,Au

Au

Lima

Salt

Phosphate, Potash

Au

Cu,Au

Pb,Zn,Cu

10 m.y.

Tr

Crustal thickness here reaches 70 km.

Many porphyry copper deposits in intrusives, age 65-4 m.y., in Peru Chile Cordilleran

Tin-Silver Province

La Paz

Sn,Ag

Au,Cu

Nitrate

Cu

Cu

Q Cu

Fe K

Tantalum, Beryllium

Cu,Au

Sn

ARGENTINA

CHILE

Santiago

Fe

Colorado R.

Negro R.

Offshore Phosphate

Offshore Phosphate

Falkland Islands

Manganese nodules

Scotia Ridge

250 m.y.

Andean Geosyncline Folding 190-50 m.y. in successive orogenies caused by subduction of Nazca Plate under South American continent.

Peru-Chile Trench

Peru-Chile Trench

Cocos Ridge Fracture Zone

Galapagos Ridge

Galapagos Islands

Nazca Plate

Nazca Ridge

Challenger Fracture Zone

Antarctic Plate

PACIFIC OCEAN

Plate

Extensive gold, silver deposits formed during orogeny 100-65 m.y.

SALVADOR

COSTA RICA

PANAMA

Bauxite

Au

Cu

NICARAGUA

Au

5-10 m.y.

T Cu

Q

Au Q

Caribbean Plate

CARIBBEAN SEA

JAMAICA

DOMINICAN REPUBLIC

PUERTO RICO

Cu

Au

Puerto Rico Trench

Lesser Antilles

Barbados

Eruption 1902

Volcanic Arc

Eruption 1979

ATLANTIC OCEAN

around which younger formations have been deposited and, in some cases, contorted by movements between these cratons. From north to south the cratons are:-

(1) The Guyana craton lying north of the River Amazon.

(2) The Guapore craton immediately south of the Amazon.

(3) The San Francisco craton that lies between the east coast and the Guapore craton, but separated from the latter in the south by a complex wedge-shaped area north and west of Brasilia.

A fourth craton, not obvious when looking at the map as it is covered to a large extent by much younger sediments and lavas, lies south of the last two and its existence is known more from geophysical evidence than from observed geology.

Each of these cratons has a nucleus of Archaean rocks (older than 2,500 m.y.) surrounded by later Precambrian sediments and volcanics which were deposited over and around these nuclei and subjected to subsequent orogenies.

During the whole period of the Precambrian and Palaeozoic eras, according to abundant evidence from different geological features, the African continent was attached to South America. The edges of the respective continental shelves, rather than the present coast lines, were in contact with the eastern bulge of South America fitting into the embayment in the west African coast. Parting first occurred at about 180 m.y.

The western edge of the continent, at least since late Precambrian times, has been dominated by successions of sedimentation in north-south trench systems and orogenies with the same orientation. The most extensive and violent orogeny, the Andean, started at about 180 m.y. reached its peak about 100 m.y. and is still in its dying stages.

The following is a chronological summary of the major geological events, and formation of mineral concentrations, that make up the history of the sub-continent.

Pre-2,500 m.y.

Volcanic deposits, and subsidiary sediments (mainly of material derived from the breaking down of the volcanics) are found in the Guyana, Guapore and San Francisco cratons in belts separated by broader areas of gneisses and granites. The latter form the evidence of an orogeny that ended about 2,600 m.y. The atmosphere at this time is believed to have had little oxygen.

The mineral deposits formed during this period include important **gold** deposits in Brazil (e.g. the Morro Velho Mine) and in Guyana. Some of the Brazilian **iron** deposits were formed at this stage but the major production is from later formations.

2,500 to 1,800 m.y.

Early in this period the atmosphere is believed to have become oxygenated and sediments characteristic of more recent geological history began to be deposited, together with some volcanics, over and around the pre-2,500, m.y. nuclei. These were folded and altered in an orogeny, known as the "Trans-Amazonian" and which started about 2,000 m.y., in eastern Guyana and somewhat later in the other cratons but much obscured by later folding and metamorphism. It probably continued until about 1,800 m.y.

The important **iron** deposits, e.g. in the Belo Horizonte area of Brazil and in Venezuela, were initially formed in this period as very extensive but relatively low-grade iron formations in which local enrichment to high-grade ore occurred later.

1,800 to 570 m.y.

Extensive sedimentary and some volcanic deposits were laid down on and around the cratons referred to. The Precambrian formations seen in the west of the continent, and involved in the much later Cordilleran orogeny, were probably formed during this period.

An orogeny varying in time from 1,400 to 900 m.y. in different parts of the continent affected the sediments and volcanics deposited since 1,800 m.y. and also involved the basements on which they lay. A period of marine sedimentation overlapped the latter orogeny in the eastern and southern part of the continent. Recent discoveries of **lead** and **zinc** mineralisation of potential economic extent and grade have been made in the Bambui sediments of about 900 to 700 m.y. age north, south and east of Brasilia.

Another orogeny, known as the "Brazilian" took place in late Precambrian—early Palaeozoic time (varying between 800 and 450 m.y. in different areas but averaging about 600 m.y.) especially in the eastern "bulge" of the continent. This is approximately coincident with an orogeny that affected parts of Africa, termed the "Pan-African Event"—an example of the shared geological history of the formerly connected continents.

Uranium occurs in several forms of deposits in Middle to Late Precambrian of Brazil. The extensive areas in northern Brazil, southern Venezuela and the Guyanas, where late Precambrian sediments overlie older granite and gneiss basement, provide potential for uranium deposits of the type found in Saskatchewan, Canada and Northern Territories, Australia.

Apatite, the primary mineral of **phosphorus,** and minor quantities of uranium-rich minerals are mined and concentrated from intrusive carbonatites and related intrusives of late Precambrian age in Brazil.

570 to 225 m.y.

The Brazilian orogeny, as mentioned above, extended into the Palaeozoic era affecting especially the eastern "bulge" of the continent which probably remained high ground. West of this, a north to south series of basins were flooded by advancing seas which deposited sandstones, shales and limestones containing abundant evidence of invertebrate life. The seas continued to expand, especially up the east-west depression that later became the Amazon valley and in the trough along the western margin that was later to become the Cordilleran range. In the later stages of this time interval a large proportion of the continent was under seawater, at least intermittently. The sediments and lavas deposited are indicated on the map by dark and light blue, representing early and late Palaeozoic, but much of their extent is hidden under late sediments.

No mountain building occurred in South America corresponding to the Appalachian folding in North America although local folding occurred between the individual Precambrian cratons as a result of relative movement between them. Granites are found in Argentina that give dates between 400 and 300 m.y. indicating a local orogeny in that area. In the trough along the west coast, where a thickness of between 10 and 15 km of sediments accumulated, some folding probably started in the late part of this time interval but was only a precursor of the much more violent Andean orogeny.

There are few metallic deposits that can definitely be allocated to this period although some of the mineralisation in the very productive Cordilleran belt may have been remobilised from original concentrations of Palaeozoic age. Some deposits of "red bed" **copper,** with accompanying **uranium,** in the central and eastern Andes of western Argentina are in continental sediments dating from 280 m.y. onwards. Sediments of this age in South America have not been very productive of oil and gas so far, but exploration in the Cordilleran foothills of Peru and Ecuador may show Palaeozoic source beds.

An ice age occurred in the southern part of the continent beginning about 300 m.y. ago. The same event affected southern Africa, India and part of Australia, all of which were united in Gondwana, the southern part of Pangaea.

225 to 65 m.y.

Seas withdrew from much of the continent after 225 m.y. but advanced again about 190 m.y. and deposited sediments over a large part of central and eastern South America. About 125 m.y. there was a period of extensive plateau lava outpouring that covered a huge area in the Parana Basin in southern Brazil, Paraguay and Argentina between Latitudes 17°S and 30°S. This lava flooding was probably related to the splitting apart of South America from western Africa, and the inception of the Atlantic Ocean, which began at about 180 m.y. starting in the south and opening northwards with time. The parting caused the South American Plate to move westwards against the Pacific Plate and its subsidiary Nazca and Cocos Plates (see Figure 3A).

As mentioned above, an extensive trench along the west side of the continent had permitted great thicknesses of sediments, and some volcanics, to accumulate since at least 640 m.y. and some folding and intrusion had occurred in several successive periods along the same axis. The Andean orogeny started with new sediment and volcanic accumulation from about 250 m.y. on basements that varied in age from about 640 to 300 m.y. The Andean folding and intrusion was concentrated in successive pulses dated (at least in the central Andes of Chile) at about 180, 150, 120, 100 m.y., respectively. The 100 m.y. pulse was the culmination of the Andean orogeny and later, but less violent, pulses continued into the following time interval. Each pulse affected a rather narrow, sharply defined, north trending belt with its own characteristic association of rocks and mineral deposits.

A second belt of folding, largely obscured by later sedimentary cover, extends from the Cordilleran belt at a point north of Santiago southeasterly to the coast of Buenos Aires.

The larger number of ore deposits in the phenomenally pro-ductive Cordilleran mineral belt were formed after 65 m.y. The most important exception is the area of **tin-tungsten** and **tin-silver** deposits of Bolivia extending short distances north and south into Peru and Argentina respectively. These are dominantly 211-160 m.y. in age although some were formed in a much later episode, between 60 and 23 m.y. This is the second most important tin producing area of the world. Other ore deposits formed prior to 65 m.y. include some of the "Manto" base metal deposits, which range in age from 150 to 100 m.y. and stratiform manganese in volcanic-sedimentary formation of about 120 m.y. While some of the porphyry copper deposits in the Andes may have been instigated in this time interval, the great majority are dated subsequent to 65 m.y.

Exploration for **oil** and **gas** in the eastern foothills of the Cordillera—especially in Ecuador—is resulting in still modest but increasing production and reserves from sediments dating between 190 and 165 m.y. although some may have migrated from source beds of Palaeozoic age.

65 m.y. to Present

The peak of activity of the Andean orogeny had been passed by this time but there was still considerable faulting, intru-sion and periodic volcanic activity over large areas, per-sisting in some parts to the present day. The intense folding of the previous period had resulted in great thickening of the continental crust under the Cordillera, reaching a thickness of 70 km in some places. Because the continental crust is "floating" on the denser mantle, the surface of such a thickened block tends to rise relative to sea level and keeps rising as erosion removes weight from the higher altitudes. This has taken place to a marked degree in the Andes, and one of Charles Darwin's observations on his voyage of *The Beagle* was of marine shells lying on mountains thousands of metres above sea level. The uplift was accompanied by much normal faulting (and resultant earthquakes) and rapid erosion that has given the very precipitous slopes of the Andes.

Sediments, partly marine, partly freshwater, continued to be deposited in the broad, roughly north-south central belt of the continent (with a branch down the Amazon valley). These sediments, undisturbed except where close to the flanks of the Andes, have continued to be deposited until quite recent times and are shown on the map by the broad yellow band extending nearly the full length of the continent.

Volanic activity was almost continuous, in one part or another of the Andes, from the beginning of this time inter-val until 10 m.y. and periodically since then, resulting in large areas of lavas and ash beds.

The Cordilleran belt is remarkably prolific in deposits of base and precious metals and the greater proportion of these were formed after 65 m.y. Porphyry **copper**, a class of mineralisation that has provided an increasing proportion of the world's production of the metal over the past 20 years, occurs in intrusive stocks in a belt that runs close to the Pacific coast from southern Chile to northern Peru and beyond. These are particularly abundant in Chile (which in 1979 had the second largest production in the world) and vary in age from 50 to 4 m.y. The two largest such deposits in the world are Chuquicamata (33 m.y.) and El Teniente (5 m.y.), El Salvador (39.6 m.y.), El Abna (34.5 m.y.) and Rio Blanco (4.3 m.y.) are smaller deposits of the same type.

A parallel belt, adjoining the copper belt to the east, carries, in sediments and volcanics, a large proportion of the **lead-zinc** deposits of Peru and Chile, most of which appear to have been formed subsequent to 65 m.y. The **tin-tungsten** and **tin-silver** deposits of Bolivia, as mentioned in the previous section, were deposited mainly before 60 m.y. but some were formed in later episodes of intrusion and volcanic activity, between 60 and 22 m.y. ago. Smaller deposits of **gold** and **silver** were also formed in this period. Deposits of economic importance of **iron** were formed between 50 m.y. and the present at a number of locations along the length of the Andes.

Deposits of the most recent age in South America are those formed by river or stream action, e.g. the **gold** placers which were worked before the arrival of the Europeans and metal concentrations resulting from the action of surface weather-ing under tropical or sub-tropical conditions. The most im-portant of the latter are those of **bauxite** where the action of such agencies has occurred on rocks with a relatively high aluminium content. Bauxite has been known and mined in Guyana and Surinam for many years and recent exploration has proved or indicated very large reserves in Brazil on both sides of the Amazon valley. **Nickel** laterites, formed over mafic rocks by rather similar processes, occur in several parts of the continent, including Brazil and Colombia, but have not yet reached production.

In the non-metallics field, **coal** and **lignite** beds were formed between 65 and 30 m.y. in continental sediments, particu-larly on the eastern flanks of the Cordillera as mined in Colombia and Argentina and also in the northern part of Venezuela.

The two main **oil** fields in Venezuela, on each side of Lake Maracaibo, produce mainly from post-65 m.y. sediments separated by a ridge of older rocks. Further east is the Orinoco Basin which contains large quantities of heavy oil in porous sediments ranging in age from pre-280 m.y. to about 15 m.y. Recovery of the heavy oil is a major problem but, in view of the large resources, is under active study and development. Other areas of production from Tertiary sediments are in Colombia, northern Ecuador and a small part of the coast of Peru. Some production from the Cordilleran foothills of the Oriente (Peru) is from Tertiary sediments but the main source-beds are pre-65 m.y. There is some oil production from coastal and off-shore Brazil at about Latitude 20°S and from eastern coastal Argentina.

CENTRAL AMERICA

The term "Central America" is taken here to include all of the mainland between the southern border of Mexico and the Panama/Colombia border and the islands of the Caribbean Sea. The oldest rocks in the area are found in an east-west belt through northern Guatemala, Republic of Honduras and Nicaragua and in a relatively small "window" in Belize. In these areas Palaeozoic rocks predominate with some late Precambrian in Honduras. Elsewhere in the area rocks are, with very few exceptions, younger than 200 m.y.

Some time after 200 m.y. a lobe of the Pacific Plate (or its subsidiary the Cocos Plate) pushed into the America Plate between the north and south continental masses. This lobe has been named the "Caribbean Plate" and is underlain by crust that is up to 15 km thick in contrast to the 5 to 8 km thickness that is more typical of the oceanic crust. The present boundaries of the Caribbean Plate have persisted since about 60 m.y.

The Greater Antilles, on the northern edge of the plate, began to develop mainly on oceanic basement, in Jurassic (180-135 m.y.) time and volcanic activity ceased at about 60 m.y. This northern boundary now follows the Cayman and Puerto Rico trenches and is the locus of left-lateral offset and of major earthquakes.

The southern boundary, less well-defined since much of it is now welded to the South American continent, has right-lateral movement on a zone of faulting that extends from Trinidad, through northern Venezuela, and into the Andes of Colombia.

The east boundary is formed by the volcanic arc of the Lesser Antilles. This is an active arc that started with major vulcanism about 200 m.y. ago and continues to the present. Most of the rocks exposed at surface are younger than 50 m.y. The arc lies over the west-dipping subduction zone where the oceanic crust of the American Plate plunges below the Caribbean Plate. The individual islands are largely volcanic but in some cases, e.g. Barbados and Bahamas, they are capped by coral reefs or bedded material of coralline derivation. A violent volcanic eruption occurred on Martinique, and a less violent one on St. Vincent, in 1902 but since then there has only been minor volcanic activity in the Lesser Antilles

The western edge of the Caribbean Plate is formed by an east-dipping subduction zone where the Cocos/Nazca Plates plunge under it producing active volcanic arcs.

Mineral deposits are widespread in the Central American mainland and in the Greater Antilles but almost absent in the Lesser Antilles island arc. On the mainland there is a rough zoning of mineral deposits from **lead**-rich in the northwest to **copper**-rich and **lead**-poor in the south and east with some increase in **silver** and **gold** in the same direction. Thus, in northwestern Guatemala, in the belt of mainly pre-225 m.y. rocks, are several small lead-zinc mines with low silver. To the southeast, in the Republic of Honduras, are two mines (El Mochino and El Rosario) of silver-lead-zinc ore in limestones. Further southeast, in younger and dominantly volcanic rocks, are gold vein-type deposits that were known before the arrival of Europeans and acted as one of the main spurs for the early exploration. Also in this general area are copper deposits including the former Rosita mine in post-200 m.y. sediments in Nicaragua. In southern Central America there is almost no lead production but there are important copper deposits of the "porphyry" type in quite young intrusives, with Panama's Cerro Colorado being the one of largest tonnage where the mineralisation is only 5 m.y. old.

In the Caribbean, the Greater Antilles contain quite numerous occurrences of copper and gold (mostly in small deposits) but almost no lead. The largest gold-bearing body is the 30 million tonne Pueblo Viejo deposit in the Dominican Republic. There are many occurrences of the porphyry copper type but the only one of economic significance is the 200 million tonne Rio Vivi-Tanama zone in Puerto Rico.

In both the mainland and Greater Antilles the most recent metallic deposits (formed in the last two million years) are those of **nickel** laterites and **bauxites**, both formed by residual concentration and re-cycling of nickel and aluminium respectively, by the action of surface waters under tropical or sub-tropical conditions. Nickel laterite is formed on relatively nickel-rich (but uneconomic) mafic rocks, often of ocean floor origin pushed up during tectonic plate movements. Surface weathering tends to remove the non-metallic constituents and to leave iron and nickel concentrated in a surface laterite. There tends to be an even higher concentration of nickel, from solution and redeposition, at the lower boundary of the laterite with the underlying primary rock. The most important nickel laterites in Central America are in Guatemala, Cuba and the Dominican Republic.

Bauxite is formed by a related, but somewhat different, process acting on the surface of rocks or beds relatively alumina-rich and iron-poor. The most important deposits in Central America are in Jamaica and Hispaniola Island (Haiti and the Dominican Republic).

Within the Caribbean Plate, accumulations of **oil** and **gas** have taken place in sediments ranging in age from 200 to 30 m.y. Concentrations of economic importance have been found off the Venezuelan coast (the reserves and production of inland Venezuela are referred to in the notes of South America) and offshore Trinidad.

In a lake on the island of Trinidad itself is a rather unique concentration of **asphalt,** the most viscous and heavy form of petroleum.

In a global framework, the area of this map sheet forms the northwestern part of the Eurasian tectonic plate which extends from the Mid-Atlantic Ridge to the Pacific coast of Asia and from the North Polar regions (where these two boundaries converge) to the southern plate boundary which passes through the Mediterranean Sea and easterly through the northern Indian Ocean. In more ancient geological history, prior to 1,800 m.y. ago, northwestern Scandinavia, Scotland and Ireland were part of the North American continent. From 1,000 m.y. or earlier, the American and European continents were separated by a trough, or combination of troughs, of varying widths which might be regarded as a "proto-Atlantic Ocean". In this system of troughs, sediments and volcanics accumulated and were compressed by repeated convergences of the sides which finally formed, in a single continuous structure, the Appalachian belt of folding in North America and the corresponding Caledonian mountains of northwest Europe. From 400 to 100 m.y., the European and American continents were sealed together along this structure.

At about the same time as the orogeny that produced the Caledonian and Appalachian mountain system, a somewhat similar, but less extensive orogeny took place in the east, beyond the limits of this map sheet. Here a north-trending trough separated Siberia from Europe until, here too, convergence of the two sides folded the accumulation of sediments and volcanics to form the Ural Mountains and seal the formerly separated plates into one.

To the south of this sheet, central and southern Europe underwent a long period of recurrent folding due to the convergence and rotation of the African plate relative to the Eurasian but this had major effects only on the southern edge of the area covered by this particular map.

The separation of the Eurasian Plate from the North American, to which it had been attached continuously since about 400 m.y., started about 100 m.y. This was later than the first Atlantic rifting between Africa and North America which started at about 180 m.y. The actual separation of Greenland from Canada occurred about 80 m.y. and continued until 60 m.y. when the parting of Ireland-Scotland-Scandinavia from Greenland commenced and still continues at present.

Pre-2,500 m.y.

The history of this period is clouded in northwest Europe by the effects on structure, rock composition and radioactive dates by later events. It appears, however, that the rocks forming the Outer Hebrides Islands of northwest Britain and on the Norwegian coast, may be regarded as part of the older nucleus of the North American craton. Larger areas of pre-2,500 m.y. rocks are found in southern Finland. The period ended with a major orogeny about 2,500 m.y. corresponding to that which occurred in many other parts of the world. No important non-ferrous mineral deposits in northwest Europe are proved to date back to this period. Banded **iron** formation occurs in the Sydvaranger area adjoining the USSR boundary. Here iron is mined in a sedimentary-volcanic series that gives dates between 2,900 and 2,500 m.y.

2,500 to 1,700 m.y.

Following the mountain building of 2,500 m.y. mentioned above, a period of sedimentation and volcanic activity ensued for the next 700 m.y. or so, probably with some periodic folding and intrusion. The effects of this are seen particularly in eastern Sweden and southern Finland where belts of volcanic and sedimentary rocks are separated by wide areas of gneiss and intrusive granites. These give dates of between 2,300 and 1,900 m.y. evidently representing periodic orogenies that probably ended about 1,800 m.y.

This period is one of the most productive of ore deposits in northwest Europe. In southern Finland, a composite belt trending northwesterly contains the important **copper, nickel, zinc** and **cobalt** deposits mined at Outokumpu and other deposits of copper, zinc and lead in the same general area. There is a gradation in age from the east part of this belt dating about 2,300 m.y. to the west where ages as young as 1,800 m.y. are obtained. Across the Gulf of Bothnia in Sweden rather similar mineralisation dating about 1,900 m.y. old is found in the Skellefte district where mining of **copper** and other base metals has been carried out in the vicinity of Boliden continuously over several centuries.

An additional area of base metal and some precious metal deposits of the same general age is found in southern

Sweden in the vicinity of Bergslagen, which also has a long mining history.

In northern Sweden is the Kiruna iron province, associated with volcanic rocks older then 1,900 m.y., which is the site of previous important **iron** mining activity.

1,700 to 570 m.y.

There is rather little evidence on the geological history in northwest Europe of the period following the orogeny of about 1,800 m.y. It is known that another period of folding and granitic intrusion occurred about 950 m.y. especially in southern Norway where important deposits of ilmenite (ore of **titanium**) are mined and also some minor **nickel** and **molybdenum**.

The most recent Precambrian, between 900 and 570 m.y. is characterised by sedimentary rocks lying unconformably on the older basement. Widespread tillites in Scandinavia and Scotland show that an ice age occurred during part of this time interval. In northwest Ireland, the Highlands of Scotland and coastal Scandinavia, the sediments of this period were involved in the later Caledonian (400 m.y.) orogeny but, on the southeastern flanks of these mountains in Sweden, the late Precambrian beds are almost undisturbed.

570 to 400 m.y.

In the early Palaeozoic, seas probably covered the greater part of the present land surface of this map-sheet and continued to do so over much of this period. Deposition of shales, limestones and sandstones coincided with the rapid development of marine invertebrate life. The marine sedimentation was interrupted at times by volcanic activity, the effects of which are seen especially in Britain, in the period between 500 and 440 m.y.

Deposits of **copper** and **lead** in Wales, northwest England and eastern Ireland (Avoca) were formed in this period. These were probably related to volcanic activity although it has been suggested that some metals may have been remobilised to their present host rocks from underlying late Precambrian formations. Although these deposits were not large by present standards, they were of historic significance since it was to gain access to these sources of copper and lead that the Roman legions marched out of their settlements in southern and eastern England and undertook the military conquest of Wales and northern England.

Base metals were also deposited in Scandinavia during this geological period and, in most cases, have been remobilised in the subsequent Caledonian orogeny.

The deposition of sediments and volcanics was interrupted, from 500 m.y. onwards, by several pulses of folding and intrusion. The earliest of these followed geographically the general trend of the present English Channel (La Manche) starting in late Precambrian time and ending about 550 m.y. Granites dating about 550-450 m.y. are found on the Brittany and Normandy coasts and in the Channel Islands. Granites of approximately the same age occur in the Ox Mountains of western Ireland and in the Scottish Highlands. Much more extensive were the effects of the Caledonian orogeny which started about 450 m.y. and reached its peak about 400 m.y. This affected all the rocks and mineral deposits formed up to this time in all of Scotland, northern and western England, Wales and Ireland as well as Norway and western Sweden. The orientation of the folding was northeasterly (on present geography) corresponding approximately to that of the major trench system that separated North America from Europe, in which sediments and volcanics had accumulated.

Many of the important **non-ferrous base metal** deposits of Norway and Sweden are in rocks affected by the Caledonian orogeny but it is difficult to distinguish those initially formed during the orogeny from those of earlier origin but altered and remobilised by it.

400 to 260 m.y.

As in most orogenies, the thickening of the crust by folding resulted in a general uplift of the areas affected. This was followed by rapid erosion and deposition of sand and gravel, much of it in fresh-water bodies, forming the "Old Red Sandstone" of Britain. The land subsided again and seas invaded a large proportion of the areas covered by this map-sheet depositing shales and limestones.

The period from about 345 to 260 m.y. was one in which the most important deposits of **iron, lead, copper, tin** and **coal** were formed in Britain and Ireland which allowed this relatively small group of islands to reach a dominant position at the time of the Industrial Revolution. The oldest of these deposits were formed at the base of the Carboniferous sediments, about 345 m.y., when seas were invading the Devonian surface. **Zinc, lead** and minor **copper** bodies were formed in bedded deposits, especially in Ireland where three zinc-lead mines are currently in production, including the Navan mine 60 km northwest of Dublin which is the largest known concentration of zinc in Europe. Some lead and zinc were deposited in beds of the same age in England but here the most productive concentrations have resulted from the remobilisation of the metals into veins in later sediments.

After a period of quiet marine sedimentation over much of northwest Europe, elevation took place raising Britain and much of Europe just above sea level and exposing areas of flat topography with many shallow freshwater bodies and swamps. A tropical climate, abundant vegetation (especially tree-ferns), and its periodic burial under clay and sand as the basins were submerged under shallow water, made ideal conditions for **coal** formation. The extensive coal basins of Britain and on the European mainland, were formed during the period between 325 and 290 m.y.

During this non-marine sedimentation, the next orogeny, the Hercynian, began about 280 m.y. and continued at its maximum activity until about 255 m.y. Its major effects were in southern Europe since the cause of the orogeny appears to have been the convergence and collision of the African and Eurasian Plates. There was, however, considerable folding in southern Ireland and southwest England and abundant granitic intrusion in the latter. There was also volcanic activity at about the same time in the area of the Scottish Midlands.

The vein deposits of **tin** and **copper** in Cornwall and Devon are associated with the granitic intrusions there. These ore deposits, although modest in size by modern standards, played a vital part in human history. They were the major source of tin which was alloyed with copper to make the harder bronze used in the Bronze Age and subsequently Cornwall maintained its position as the leading producer of tin for many centuries and, during a period in the 18th Century, southwest England was also the leading world producer of copper.

Athough rather few ore deposits of importance appear to have originated in Scandinavia at the time of this orogeny, some metallic concentrations that had been formed earlier may have been remobilised into new positions and forms during the folding and intrusion to the south.

260 m.y. to Present

During the Hercynian orogeny in the early part of the Permian period, there were two main basins in the North Sea area in which non-marine and marine sediments accumulated, a northern and a southern, separated by a "high" corresponding approximately with the North Sea

"high" and the Ringkobing-Fyn "high" marked on the map. There was some volcanic activity, mainly submarine, remnants of which are now exposed in southwest England and mid-Scotland. The climate by this time was arid and, in the later stages of the Permian, arms of the sea were cut off resulting in increased salinity and deposition of salt, gypsum and potash. The **potash** deposits of Germany, for many years the world's largest supplier, were formed about 240 m.y. as were those at depth under the east coast of England.

High grade iron deposits, in the form of pod-shaped haematite bodies were formerly mined in northwest England west of the Lake District. They occur at and below the unconformity of basal Carboniferous lying on older Palaeozoic but appear to have been formed sometime after 180 m.y. Collectively these deposits have produced over 150 million tonnes of high grade ore and, with the larger tonnages of low-grade ores from the Coal Measures, played an important part in the early stages of the Industrial Revolution. Before smelting with coal was achieved, the haematite ores were transported by pack horses from the coastal regions over the hills to valleys where there was sufficient timber to provide charcoal for smelting. Most of these deposits are now mined out and their place in the steel industry was taken first by Jurassic bedded iron (160-135 m.y.) and later by imported ores.

Another formation of economic interest in the upper Permian, about 250 m.y. is a remarkably persistent **copper**-bearing bed in northern and northwestern Europe. This was first worked in Germany about the 14th Century under the name "Kupferschiefer" and has subsequently been followed at depth into Poland. The same formation can be traced northwesterly to the east coast of England where it still has above-average metal content but is not of economic significance.

As the following Triassic period proceeded, sedimentation was still dominantly non-marine over northwest Europe. On the shores of these inland bodies of water were seen the expansion in numbers and diversity of dinosaurs which became the dominant form of land vertebrate life for the next hundred million years.

It was approximately 180 m.y. that the break-up of Pangaea commenced (see Figure 5, p13). Although of immense significance on a global basis there was little direct effect on the pattern of sedimentary and volcanic deposition in northwest Europe since this did not break away from Greenland until 60 m.y. Possibly related to the initial break-up of Pangaea, however, was the inception, in the area of the present North Sea, of the tectonic pattern which controlled the subsequent history of this area and the development of the **hydrocarbon** deposits now being exploited. The pattern that developed was one of deep, linear, sediment-filled troughs, up to 50 km wide and 320 km long, separated by fault-bounded platforms of continental crust. These began to develop after the Hercynian orogeny, about 250 m.y., and the vertical movement did not become dormant until about 100 m.y. The process was one of graben faulting, i.e., wedges of rock let down between parallel faults. The major structural units are the Viking and Central Grabens (shown on the map) which extend southerly to the southern North Sea Basin. A branch graben extends westerly to the Moray Firth Basin, off northeast Scotland. Some similar structures appear to exist west of Scotland but have not been explored in detail.

The deposition of subsequent sediments was, naturally, thicker in the graben sub-sea valleys than on the flanking ridges and, as seen on the map, the majority of oil and gas pools are concentrated in these sediment-filled troughs. The seas retreated to some extent after the initial graben formation, allowing partial erosion of the earlier sediments, and then, at about 150 m.y. advanced again depositing unconformably the Kimmeridge bituminous clay which was the major source-bed for most of the oil that accumulated in the overlying younger porous beds. Some additional block-faulting occurred which helped to form traps containing oil and gas reserves.

In Middle Cretaceous, about 100 m.y. there was further transgression of seas both from the precursor of the Mediterranean to the south and from the Boreal Sea to the north. At this time, much of northern Europe became submerged and the North Sea Basin, centred on the Viking Graben, continued to subside and to permit thick sedimentation.

At the end of the Cretaceous (65 m.y.), the Tertiary started with general uplift, and consequent erosion, and only the centre of the North Sea Basin remained under water. This was followed by renewed subsidence to the extent of 3,500 metres in the centre (but with some temporary reversals) which has led to the present North Sea. Volcanic activity occurred, about 55 m.y., in the North Sea but still more near the Atlantic coasts of Scotland and Ireland (the Giants Causeway) and also in coastal Greenland which was then quite close.

By this time, most of the graben troughs of the North Sea pattern had been filled and the Tertiary sediments were deposited fairly evenly over large areas in gentle basins that extended onto areas that were now land in eastern England and northern and western continental Europe. Most of Norway, Sweden and Finland remained above sea level throughout this period.

The genesis of the North Sea oil field, the most important discovered in the last quarter century, took place mainly between 220 and 30 m.y. although some of the source beds, from which the hydrocarbons were derived, date back to the Coal Measures of about 320 m.y. As mentioned in the Introduction, the source of most oil and gas is simple and minute forms of marine animal life that are embedded in bituminous clays. Under the weight and pressure of superimposed sediments, and the resulting rise in temperature, the decomposing organic material is changed to petroleum products which migrate into more porous, and usually overlying, formations.

The main traps for oil and gas that developed in the North Sea field are (according to K. Skipper):-

(1) Fractured and porous Permian beds (average 260 m.y.), under southern British and Dutch waters, in anticlines capped by salt beds which trapped gas migrating from the Carboniferous Coal Measures.

(2) In the Ekofisk Province, oil and gas has accumulated in anticlines in 90 to 50 m.y. beds at depths of 3,300 to 4,000 metres.

(3) In the central North Sea, e.g. the "Forties", reserves are in sandstones of about 65 to 55 m.y. at depths of 2,400 metres.

(4) In the northern Brent Province accumulations occur in fault blocks of Jurassic sandstone (average 170 m.y.) with a thickness of about 300 metres.

(5) On the western edge of the Shetland Platform accumulations of heavy oil have been proved in 500 metre-thick sandstones of about 200 m.y.

The recoverable reserves of oil in the North Sea fields were estimated, in 1979, to be between 18 and 20 billion barrels (2.6 billion tonnes) of which 12 billion barrels were in the United Kingdom sector. On the basis of the above figures the North Sea fields would constitute between 2.5 and 3 per cent of known world reserves.

Natural gas reserves are more difficult to establish but those of the North Sea, including the Groningen field that is mainly under land in the Netherlands, are estimated to be over 110 trillion cubit feet (3.1 trillion cubic metres) which is about 4 per cent of reported world reserves.

Other oil and gas reserves within the area covered by the Northern Europe map sheet include a Triassic (about 200 m.y.) gas field off Blackpool on the west coast of northern England being prepared (1980) for production; another currently producing south of the southern coast of Ireland; and the only on-shore oil-field in Britain which is in Triassic beds below Corfe Castle/Wareham on the south coast of England.

A mineral resource of importance, especially in the 19th and early 20th centuries, is bedded **iron** laid down between 200 and 155 m.y. Although low-grade (about 28% Fe) by modern standards these ironstones played an important part, together with higher-grade iron ores from northwest England, in the Industrial Revolution in Britain and continental Europe, before the availability of high-grade Precambrian ores from other parts of the world.

During the sedimentation on the northwest continental shelf of Europe between 200 and 20 m.y. there was relatively little folding on the area covered by this map sheet. This was in spite of the repeated surges of activity in central and southern Europe and Asia Minor that resulted from the convergence and rotation of the African Plate. The last pulse, between 15 and 5 m.y. that formed the Alps, produced moderate folding of the Mesozoic and early to middle Tertiary beds of southern England and renewed movement on pre-existing faults in the North Sea.

The Pleistocene Ice Age

About 2 million years ago, the world underwent a general cooling in climate and ice caps began to develop over the North and South Poles. By 600,000 years ago, a continental ice sheet had covered much of northern Europe. The ice, which advanced and receded with cycles of colder and warmer weather, reached a maximum southerly line that passes approximately through London, the Hague, Dresden and Krakow. A smaller, isolated ice-cap covered the Alps.

In addition to the direct effects of glaciation in surface scouring and deposition of morainal material an indirect effect felt over the entire world was the lowering of sea level due to the locking up of water in polar ice-caps. This resulted in the exposure of land that had previously been under water. It is calculated that the sea level over the world at the maximum extent of Pleistocene glaciation (18,000 years ago) was 100 metres or so below that at present. The significance of the additional land surface, and land bridges, on the spread and development of mammalian life, including Man, will be appreciated.

It has also been calculated that if the remaining ice-caps of the world today were melted the sea level over the whole globe would rise 110 metres above that now existing and, in the area of this map sheet, would cover the low-lying parts of eastern England and the "Low Countries".

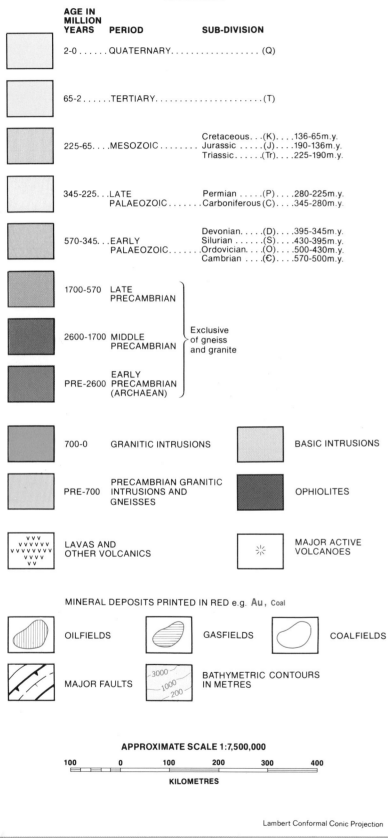

NORTHERN EUROPE

LEGEND

AGE IN MILLION YEARS	PERIOD	SUB-DIVISION
2-0	QUATERNARY	(Q)
65-2	TERTIARY	(T)
225-65	MESOZOIC	Cretaceous...(K)....136-65m.y. Jurassic.....(J)...190-136m.y. Triassic.....(Tr)...225-190m.y.
345-225	LATE PALAEOZOIC	Permian.....(P)...280-225m.y. Carboniferous(C)...345-280m.y.
570-345	EARLY PALAEOZOIC	Devonian...(D)...395-345m.y. Silurian.....(S)...430-395m.y. Ordovician...(O)...500-430m.y. Cambrian...(€)...570-500m.y.
1700-570	LATE PRECAMBRIAN	
2600-1700	MIDDLE PRECAMBRIAN	Exclusive of gneiss and granite
PRE-2600	EARLY PRECAMBRIAN (ARCHAEAN)	

700-0 GRANITIC INTRUSIONS

BASIC INTRUSIONS

PRE-700 PRECAMBRIAN GRANITIC INTRUSIONS AND GNEISSES

OPHIOLITES

LAVAS AND OTHER VOLCANICS

MAJOR ACTIVE VOLCANOES

MINERAL DEPOSITS PRINTED IN RED e.g. Au, Coal

OILFIELDS

GASFIELDS

COALFIELDS

MAJOR FAULTS

BATHYMETRIC CONTOURS IN METRES

APPROXIMATE SCALE 1:7,500,000

100   0   100   200   300   400

KILOMETRES

Lambert Conformal Conic Projection

The area covered by this map sheet is now part of the Eurasian tectonic plate extending from the mid-Atlantic Ridge in the west to the Pacific Coast of Asia and from the Arctic polar regions to the southern plate boundary along the Mediterranean. Prior to about 400 m.y. the European portion, including the area of this sheet, was separated from Asia by a trough extending roughly north-south and of unknown width. Sediments and volcanics of great thickness were deposited in this trough from about 500 m.y. until the two sides converged and collided, crumpling and overthrusting the sedimentary-volcanic accumulation to form the Ural Mountains and seal together the two plates to make the single Eurasian Plate that has remained since. The culmination of this orogeny, about 390 m.y. corresponds in time approximately with that of the Appalachian and Caledonian orogenies in North America and northwest Europe-Greenland respectively.

Previously, western Europe lay close to North America and remained so, with periodic convergences and withdrawals, from over 1,000 m.y. until 180 m.y. In late Precambrian to Cambrian times (about 600 m.y.), and again during the Caledonian orogeny (about 400 m.y.), folding and metamorphism occurred in some western and northern parts of the area of this map sheet, e.g. in the Iberian Peninsula, in the Massif Central and in Bohemia. The effects of these disturbances, however, have been largely obscured by younger sediments and by later orogenies. The cause of these later orogenies, especially active about 280, 60, 30 and 15 m.y., appears to have been the convergence and rotation of the African Plate relative to the Eurasian.

As may be seen by a glance at the map, Precambrian basement is exposed in relatively few parts of this area—in the Brittany Peninsula and the Massif Central of France; in several areas in the west of the Iberian Peninsula; in limited parts of the Bohemian Massif centred in western Czechoslovakia; in a belt from northern Greece to Yugoslavia; north of the Black Sea in USSR; and at the "toe" of Italy. These rather limited "windows" where the Precambrian rocks are not covered by later strata confine our knowledge of the history of this era more than, for example, in Scandinavia or Africa. It is probable that the Precambrian platform, on which later formations were built, has approximately the same time divisions as are found in other, better exposed, Precambrian areas. Radioactive dating in Brittany indicates a late Precambrian age, 900 to 600 m.y., but it is probable that older rocks exist but have been reworked at this later period.

On this Precambrian platform and in trench systems on its flanks, marine sediments and volcanics were deposited in stages between 570 and 280 m.y. On the southern flank, the trenches varied in extent and position but probably had a dominantly easterly trend relative to the present orientation of Europe. Marine deposition was broken by at least three periods of folding the first of which (about 500 m.y.) was much obscured by later episodes and the second, the Caledonian 400 m.y., was of much greater magnitude in northern and northwestern Europe than in the areas on this map sheet. The third orogeny, the Hercynian, which reached maximum activity from about 280 to 255 m.y. affected most of the area of this map. Its effects are seen in the folded pattern and extensive granitic intrusions in Portugal and Spain, in Brittany and southern France and in the area surrounding the Bohemian Massif. The pattern of this orogeny, which extended eastwards beyond the limits of this map sheet, has been partly obscured by the much later orogeny that formed the Alps and related ranges.

The succession of geological events, and the formation of the present mineral sources of energy and metals, are summarised chronologically below.

Precambrian to pre-570 m.y.

As mentioned above, the exposures of Precambrian rocks in central and southern Europe are too limited to derive any complete or orderly account of the history of this period. Almost certainly the Precambrian platform includes a nucleus of Archaean rocks affected by an orogeny about 2,500 m.y. but its extent is unknown. There was another major orogeny between 1,600 and 1,900 m.y. affecting the Precambrian of southern Finland and eastern Sweden.

No important mineral deposits in southern Europe can be definitely dated as Precambrian although in some mining areas, that give later dates, the metals may have been derived from original older concentrations. On the basis of Precambrian areas in other parts of Eurasia, and in other continents, it is most probable that metallic deposits of this age occur hidden below later formations.

570 to 400 m.y.

During this interval, the continental shelf off western Europe lay close to, or against, the shelf off eastern Canada and Greenland. Much of the area of this map sheet was probably submerged by seas that advanced from the south over the Precambrian platform. This resulted in the deposition of shales, sandstones and limestones, with abundant invertebrate life, interspersed in places with volcanic deposits. The greater part of these sedimentary-volcanic deposits are now hidden by later strata.

Metallic deposits were formed during this interval especially in the vicinity of volcanic activity. One of the more important metalliferous belts is that which includes the **pyrite** and **copper**, and associated **manganese** deposits of Rio Tinto, Spain, which have been worked since Roman times at least and are still producing. This mineralisation is associated with a period of volcanic activity, interspersed with sediments, dating about 450 m.y. The mineral deposits and enclosing rocks were later involved in folding, faulting and intrusion. Probably of the same approximate age are low-grade **gold** deposits in northern Spain, which were the source of placer gold deposited in geologically recent times. Some of the base metal deposits found in the Pyrenees and the Southern Alps may have been formed initially at this time but in many cases have been remobilised during later folding and intrusion.

Some **hydrocarbons** were probably formed in marine shales deposited during this time interval but would be mainly dispersed at surface during the succeeding folding, faulting and metamorphism.

A period of folding occurred at about 500 m.y. and although its extent and magnitude are not known in detail, due to the effects of subsequent orogenies, intrusive rocks dating between 500 and 400 m.y. are seen on the northern coast of Brittany and Normandy as mentioned in the notes on the Northern Europe sheet. The period ended with a major orogeny, the Caledonian, culminating about 400 m.y. This showed its most marked effects in northwestern Europe (as described in the notes to the Northern Europe sheet) but affected southern Europe to some degree, especially in the Iberian Peninsula which at this time lay closest to the east side of the North American Plate.

400 to 230 m.y.

With the decline of the Caledonian orogeny in northwest Europe there was a general uplift relative to sea level for much of Europe. This coincided with granitic intrusion which may have extended over much of central Europe but is exposed only in the Bohemian Massif where granites give dates around 350 m.y.

The uplift resulted in rapid erosion over much of Europe and this was followed by a period of arid climate. This in turn was succeeded by new incursions of the seas depositing sediments that were interlayered with lavas and ash beds in local centres of volcanic activity. It was a time of rapid growth in numbers and diversity of life in the seas including the development of fishes which, as the first form of vertebrate life, had appeared at the end of the previous period.

By about 325 m.y., much of Europe was barely above sea level with large areas of swamp. Under a tropical climate that had succeeded the arid period, and the rapid development of land plants (with contemporaneous evolution of land fauna including amphibians and insects), conditions were ideal for the initial stages of **coal** formation. Vegetation grew and died forming peat which was buried by layers of mud or sand in shallow water and became lignite. Under deeper burial this in turn became coal of various qualities depending on original composition and the extent of alteration. These accumulations took place in broad, shallow basins which were involved in later gentle folding and faulting. The coal basins in Britain are referred to in the notes on the Northern Europe sheet. On the European continent, coal basins of this age are in Belgium, northern France and Germany and through the Silesian fields of Poland to Turkey. In many cases, the coal beds are buried under younger formations.

Metallic orebodies formed during this time interval include the German (FDR) Rammelsberg **copper-zinc-lead** deposits, mined for a thousand years, and the Meggen **copper-zinc** ores, both in sediments with some volcanics, deposited between 360 and 350 m.y. ago in Middle Devonian times. Also originating in this general period are **copper** and **lead** ores in mines worked for many centuries in Czechoslovakia and Germany (DDR). Some are related to granitic intrusives dating between 370 and 325 m.y., apparently a late phase of the Caledonian orogeny. **Copper** and **lead** deposits in western Spain were formed in the same general period. Some of the **iron** mined in Turkey today, and in earlier history, is in formations of late Palaeozoic age.

The upheaval of the Hercynian orogeny, the most active part of which occurred between 280 and 250 m.y. seems to have resulted from the convergence of the northwestern corner of the African continent on the southern edge of the

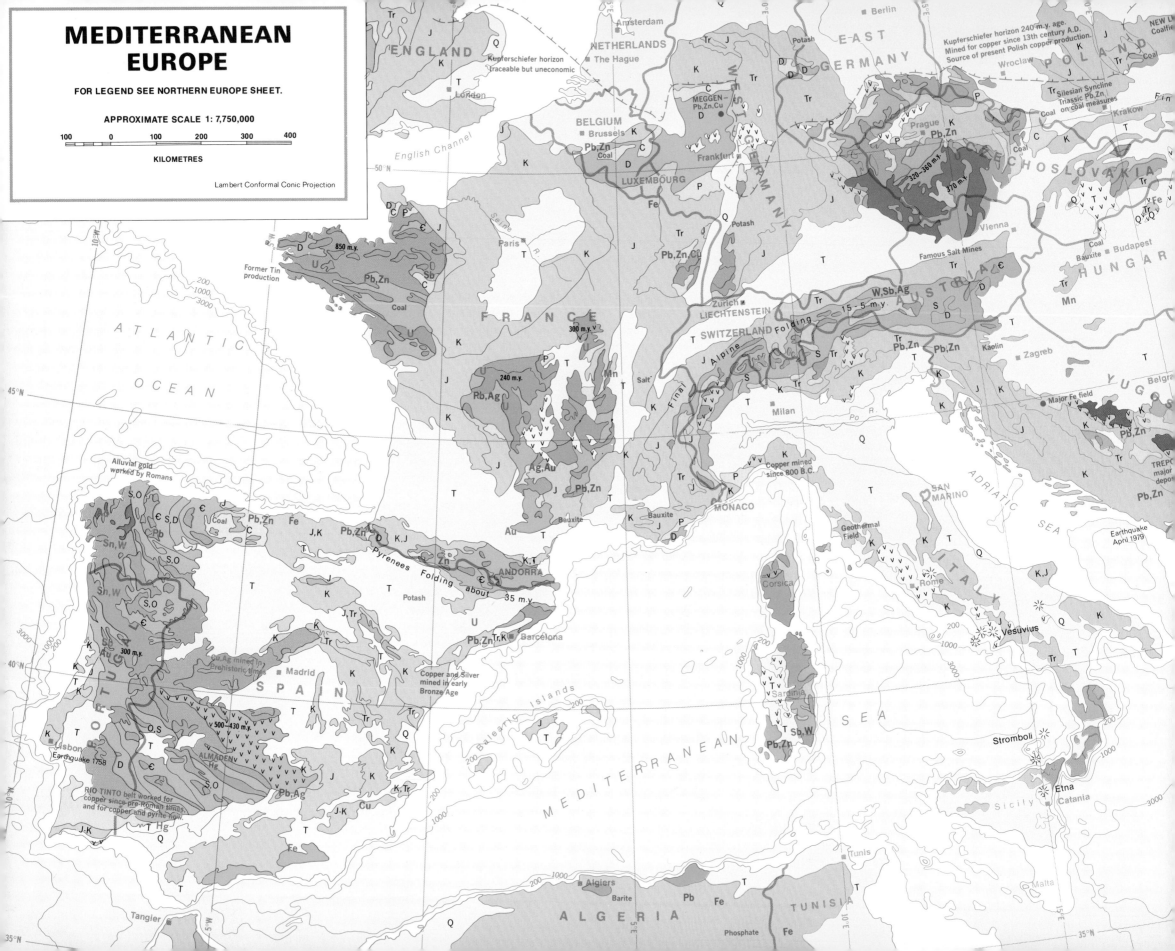

MEDITERRANEAN EUROPE

FOR LEGEND SEE NORTHERN EUROPE SHEET.

APPROXIMATE SCALE 1: 7,750,000

100   0   100   200   300   400

KILOMETRES

Lambert Conformal Conic Projection

Kupferschiefer horizon 240 m.y. age.
Mined for copper since 13th century A.D.
Source of present Polish copper production.

Silesian Syncline
Triassic Pb,Zn
on coal measures

Kupferschiefer horizon
traceable but uneconomic

Alluvial gold
worked by Romans

Former Tin
production

850 m.y.

240 m.y.

300 m.y. v

300 m.y.

Pyrenees Folding about 35 m.y.

v 500-430 m.y. v

Cu,Ag mined in
Prehistoric times

Copper and Silver
mined in early
Bronze Age

Copper mined
since 800 B.C.

Alpine Folding 15-5 m.y.

RIO TINTO belt worked for
copper since pre-Roman times
and for copper and pyrite now.

Earthquake 1758

ALMADEN

Earthquake
April 1979

Geothermal
Field

Famous Salt Mines

Countries and seas (labels on map)

ENGLAND, NETHERLANDS, BELGIUM, LUXEMBOURG, FRANCE, SPAIN, PORTUGAL, EAST GERMANY, WEST GERMANY, POLAND, CZECHOSLOVAKIA, AUSTRIA, SWITZERLAND, LIECHTENSTEIN, HUNGARY, YUGOSLAVIA, ITALY, MONACO, SAN MARINO, ANDORRA, ALGERIA, TUNISIA

ATLANTIC OCEAN, MEDITERRANEAN SEA, ADRIATIC SEA, English Channel, Balearic Islands

Cities and places

Berlin, Amsterdam, The Hague, London, Brussels, Frankfurt, Prague, Wroclaw, Krakow, Vienna, Budapest, Zagreb, Paris, Zurich, Milan, Rome, Madrid, Lisbon, Barcelona, Tangier, Algiers, Tunis, Catania, Malta, Belgrade

Potash, MEGGEN—Pb,Zn,Cu, Coal, Pb,Zn, Salt, Bauxite, Kaolin, Mn, Corsica, Sardinia, Vesuvius, Stromboli, Etna, Sicily, Major Fe field, TREPCA major deposit

Pb,Ag, Ag,Au, Sn,W, S,O, Hg, Barite, Phosphate, Fe, Cu

Seine R., Po R., Final

Eurasian Plate. This has resulted in a very complex pattern of folding, as may be seen on the map, having an average easterly orientation but varying through over 90 degrees in each direction. The ranges of hills or mountains resulting from this orogeny include the north to northwest-trending folds in Portugal and Spain and probably the first stage of the Pyrenees; the east to northeast folds from Brittany to the Ardennes and on to the Harz Mountains; the complex folding surrounding the Bohemian Massif and the Massif Central of southeastern France. The folding involved the sediments and volcanics that had been deposited on the southern flank of the European platform since 570 m.y.

In many places it is not easy to separate the effects of the Hercynian orogeny from those of the later Alpine folding. In fact there was probably no distinct break between the two. Broadly, however, the Hercynian orogeny is the event responsible for the geological pattern, topography and scenery of much of the area covered by the Mediterranean map sheet.

The latest Permian beds, dating about 250 m.y., include a remarkably extensive metal-bearing formation known as the "Kupferschiefer", that has been mined in Germany since the 13th century. More recently the **copper**-bearing bed, which includes also lead and zinc and a small proportion of uranium, has been traced under deep cover into Poland which has, as a result, become the largest copper producer in Europe. The same stratigraphic layer, still metalliferous but not of economic grade, extends westerly through the Low Countries and across the North Sea to eastern England.

Vertical movement resulted in cut-off arms of the sea with local increased salinity and deposition of **salt, gypsum** or **anhydrite** and **potash** between 250 and 230 m.y. The great German **potash** deposits, the world's largest until discoveries in western Canada, were formed at this time.

230 to 65 m.y.

At the start of this time interval, the Iberian Peninsula and the continental shelf off Britain and Ireland lay snugly against the continental shelf off the Canadian Maritime Provinces and east Greenland. Africa, the long axis of which was then in a more southwesterly direction than today, was in contact with the Eurasian tectonic plate in the vicinity of Morocco and Algeria and with North America at its western "bulge". Italy at this time was part of the African tectonic plate and a wedge of ocean, which widened to the east from Turkey, separated Arabia from Asia (see Figure 5).

As the Hercynian folding died down there occurred, as is typical of post-folding periods, a general uplift of the European area and widespread desert conditions. There was open sea in what is now the eastern Mediterranean and marine invasions extended into central Europe. There was a further period of **salt** deposition in the period 210 to 200 m.y.

The Triassic beds in Europe generally have a higher-than-average content of mineable metal an example being the **lead** and **zinc** mines found in the dolomites within the Polish Silesian.

The main **uranium** supplies of France come from the southern part in sediments of this general age and in fractures in the basement on which the sediments lie. The uranium appears to have been leached by surface waters from granites of about 300 m.y. and deposited in beds containing organic trash or in basement fractures. Uranium in sediments of about the same age (230 m.y.) is reported from Italy, Yugoslavia and Hungary.

There was increasing marine invasion as the Mesozoic era continued and sediments were deposited over much of central and southern Europe. Volcanic activity occurred in many areas of the southern part of the European Plate and also in Turkey which, at this time, was probably still part of the Eurasian Plate, the southern boundary of which is indicated by a belt of ophiolites extending from Cyprus through eastern Greece to central Europe.

Many metallic deposits were formed during this time interval in southern Europe. It is a fact that the larger proportion (by value) of metals mined in the past and currently in Europe south of the Baltic Sea, and in Anatolia (Turkey in Asia), were formed since 230 m.y. This is in contrast to northern Europe, Africa or North America where the larger proportion of metal value comes from Precambrian or early Palaeozoic rocks. It has also been noted (by Petraschek) that north of the Alpine-Carpathian mountain ranges copper tonnage (production plus known reserves) is 10 times that of lead plus zinc. In contrast, in the southern part of the fold belts, over 30 times as much lead plus zinc has been produced, or is in reserves, as copper.

In age, the mineralisation in the Alps and western Carpathians is mainly Mesozoic (230-65 m.y.), that in the Balkans and Anatolia is Upper Cretaceous to Eocene (about 80 to 50 m.y.) and the deposits in Spain (excluding the Rio Tinto belt), southern France and Italy are 37 m.y. or younger. Bedded **iron** of Mesozoic age is found in many parts of Europe.

Most of these European deposits individually are small by world standards but some have played a critical role in history. For example the **lead-silver** mines of Laurium, just south of Athens, provided the silver that formed the industrial basis of that civilisation. The **copper** and **pyrite** deposits of Cyprus, which were an important early source of the use of copper, are in ophiolite rocks of ocean bottom origin thrust onto continental crust about the same time. Also formed in the same general period were many of the deposits of **iron** and **copper** in Anatolia.

Oil and **gas** were formed in marine sediments of this time interval in several parts of Europe but are not easily separated from accumulations in later beds (post 65 m.y.) and are referred to below.

65 m.y. to Present

At the start of this period, the separation of Eurasia from North America was already in progress although the parting of northwest Europe from Greenland did not commence until 60 m.y. Africa had rotated counter-clockwise until the converging front, pushing against Eurasia, extended along the full length of the present Mediterranean. Italy and Greece were being pushed into the southern flank of Europe at one point and Turkey at another resulting in an eastern Mediterranean sea between these two and a separate sea between Italy and Gibraltar.

Sediments deposited over southern Europe at this time were partly from fresh water and partly marine, with seas sweeping in from the narrowing "Tethys Sea" between Turkey and Russia, and from the north. Volcanic activity was especially prevalent in Turkey but also occurred in the Balkans and in the Iberian Peninsula.

Metallic deposits, including **copper, lead** and **zinc**, as well as some **gold** and **silver**, continued to be formed during this period in many parts of Europe, both around the Mediterranean and further north. These include many small base and precious metal deposits in the Iberian Peninsula; the important **lead-zinc-copper** deposits of Trepca in Yugoslavia; and a sinuous, roughly south-trending chain of copper deposits, with some lead-zinc, from Romania to Bulgaria.

Some of the **copper** deposits of Anatolia were formed during this period and are of historic interest since the earliest known use of the metal, about 4500 BC, first as a colouring agent for ceramics and then for weapons or tools, was in that region. The Hittites, who occupied this area from about 2300 BC and were leaders in copper metallurgy, were also the first producers of **iron** artifacts. The ancient Egyptians did not mine or use iron, at least not until after the XVIII

dynasty, and the earliest examples were found in the tomb of Tutankhamen, buried in 1325 BC. These consisted of a miniature model of tools and a dagger evidently prized as curiosities and probably obtained from the Hittites.

Oil and **natural gas** continued to form in marine sediments that extended in time from before 65 m.y. to about 20 m.y. By world standards, no great oil fields occur in continental Europe but there are two classed as "giant oil fields" (over 500 million barrels ultimate recovery). These are near the junction of Italy, Austria and Yugoslavia and in southern Romania. Increased exploration activity in Turkey has added to reserves there and encouragement has been received in drilling off the shores of Spain, Italy and Greece.

Tectonic disturbance, resulting from movement along the boundary between the Eurasian and African Plates, occurred throughout most of this time interval. Compression from south to north started about 65 m.y. producing overfolding, or "nappes", in surges of increasing intensity up to about 35 m.y. At this time, the main Alpine orogeny resulted in the piling up, in overfolds and thrust plates, of sediments dating from 570 m.y. onwards that had already been folded in previous orogenies. The main ranges that resulted from this pressure from the south are curved, convex to the north, and include the Carpathians, where the final folding was about 30 m.y., and the Alps, the youngest and most spectacular European mountains, between 15 and 5 m.y. The Pyrenees, which had been folded in Hercynian times, were refolded about 35 m.y. ago. The most southerly fold belt of the Alpine orogeny is that passing southwesterly through southeast Spain and making a 180° turn at the Straits of Gibraltar continuing then east along the north coast of Morocco and Algeria. The folding along the length of Italy and the coast of Yugoslavia was also formed in the general Alpine period but probably rather earlier than the Alps themselves.

Anatolia, being close to the collision boundary between Africa-Arabia and Eurasia, underwent particularly violent disturbances. The compression of Arabia on Eurasia resulted in what is now Anatolia being squeezed westwards. A south-side-west movement of about 85 km is indicated along the North Anatolian Transform fault (see map) and a smaller southwesterly movement of the northwest side on the East Anatolian Transform fault. This westerly movement of Anatolia started about 15 m.y. ago and is still active to some extent.

An unusual event, probably having implications far beyond the Mediterranean, took place about six million years ago and lasted about a million years. This has been named the "Messinian Event" and included the precipitation

53

of about 1,000,000 cubic kilometres of salt and gypsum in the Mediterranean during a period when it was probably closed at intervals from outside ocean waters, allowing the addition of only about enough sea water to balance evaporation. The amount of salt that has been measured requires the evaporation of at least 2,000 metres depth of water. Before and after the Messinian Event the basin of deposition was over 2,500 metres deep. Yet the actual salt precipitation gives evidence of having taken place in shallow waters. The implication is that the sea evaporated, while precipitation was in progress, to a point where the level of the isolated body of water was far below world sea level—something like an extreme parallel to the Dead Sea. The removal of this amount of salt from the content of the oceans generally may have had significant global effects on marine life. Eventually the sea broke in from the Gibraltar end and the resulting cascades, lasting hundreds and perhaps thousands of years, must have been an imposing spectacle.

One of the last events in Europe's geological history was the general cooling of climate, about 600,000 years ago, resulting in the Pleistocene Ice Age. The main ice sheet, extending from Polar regions, barely reached the areas of this map sheet. (See notes on the Northern Europe sheet). A local ice sheet, however, covered the higher elevations of the Alps and fed glaciers that flowed down the valleys. Variations in climate, corresponding to glacial and interglacial periods, affected the position of the ice fronts and hence the conditions surrounding the life of early Man who by now had spread across Mediterranean Europe.

Since the region of the Mediterranean is the locus of the boundary, or succession of boundaries, between the African and Eurasian Plates, it is not surprising that earthquake and volcanic activity has continued up to the present. One of the most violent and catastrophic volcanic explosions, of which there is evidence in human history, occurred on the Aegean Island of Santorini in 1470 BC and destruction by the accompanying tidal waves was evidently a vital factor in the decline of the Minoan civilisation. Eruptions of Vesuvius and Etna are a part of recent history and current life. A beneficial aspect of the thermal activity is the existence of hot waters and steam, tapped in northern Italy for geothermal power, an energy source that will be increasingly used.

THE AFRICA SHEET

A glance at the map covering northern Africa and Arabia and the one covering southern Africa shows the great extent of Precambrian rocks over the continent. During most of geological history, the Arabian Peninsula was firmly attached to the African continent and only split away in very recent times, about 30 m.y. Earlier than that, prior to 180 m.y., Africa formed the nucleus of Gondwanaland (see Figure 5, p13), the southern unit of the Pangaea supercontinent and had attached to it (listing clockwise from the east side) India, Antarctica, South America and North America.

Folding of any age later than Precambrian is rarer on the African continent than any other part of the world and is restricted to the quite young (90-50 m.y.) Atlas Mountains at the extreme north, the relatively small area of folding at the southern tip of the continent (200 m.y.) and a belt on the Oman coast on the extreme east of the Arabian Peninsula. Apart from these three limited fold areas, the continuity of Precambrian rocks is broken only by an area of young basalt flows along the Rift Valley in the east and by six broad basins in which Cambrian and later sediments accumulated and were never disturbed by major folding. The six basins consist of one in the western Sahara, the next in the central Sahara (Lake Chad), and four in a line down the west-central part of the southern lobe of Africa—these have been named the Congo, Okayango, Kalahari and Karroo basins. These were structural basins during the deposition of sediments but today form quite rugged plateaux.

The time boundary between Archaean and later Precambrian was earlier (3,000 m.y. or more) in southern Africa than that of most other parts of the world (2,600 m.y.). These Archaean rocks formed centres around which later Precambrian sediments and volcanics have been built up. The result is that the African continent is built around a dozen or so cratons or platforms which have moved relative to one another during later Precambrian times, causing folding of the sediments or volcanics between them. In some cases the ancient central craton, roughly circular in plan, is clearly seen on the map—for example, the Zimbabwe craton. In others, such as the Kaapvaal to the south of the last, the cir-cular pattern is less obvious as its southern portion is covered by younger sediments.

In other cases, the original ancient nucleus is obscured by being reheated and altered by subsequent orogenies resulting in giving the dates of these later events rather than those of original formation. This is particularly so in the Nubian-Arabian craton, which must be regarded as a single block extending from Arabia to Egypt and Sudan, since this was the case prior to its very recent (30 m.y.) parting along the Red Sea fissures. A large proportion of this craton was affected by two subsequent orogenies, one about 1,000 m.y. and the other extending between 600 and 400 m.y. ago.

Below is an attempt to summarise the complex geological history of the African continent on a chronological basis.

3,350 to 2,600 m.y.

Volcanic activity (and associated sedimentation) now seen in belts separated by gneiss and granite occurred in cratons like the Kaapvaal of southern Africa, where the latest granite intrusions are dated at 3,100 m.y., and the Zimbabwe craton, where the orogeny ended about 2,700 m.y. Sediments in the Limpopo and Zambesi fold belts date about 3,000 m.y. Cratons in northern and central Africa probably had similar early histories, the results of which have been largely obscured by later orogenies, but some 3,000 m.y. dates have been noted in the western Sahara and west Africa. The main mineral deposits associated with this period are:—

(1) **Gold** in veins cutting volcanic rocks, or in fractures in iron formations, is found in nearly all pre-2,600 m.y. areas. It is usually in relatively small deposits but provided the source of gold for the later Witwatersrand gold field.

(2) **Nickel**, especially in Zimbabwe, with some associated platinum, in basic and ultrabasic volcanics.

(3) **Chromium** associated with ultrabasic lavas or intrusives—some of the high quality deposits of Zimbabwe are in rocks of this type and age.

(4) **Asbestos** in basic intrusive rocks.

(5) **Iron** in widespread but relatively low-grade iron formations.

(6) **Lithium, beryllium** and associated rare metals in pegmatite bodies.

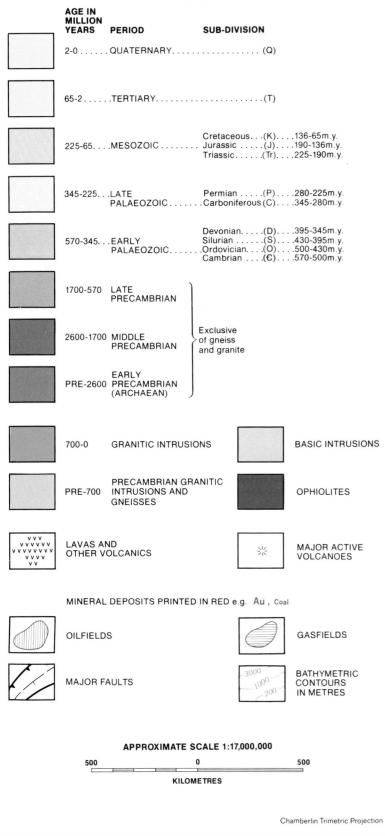

GREECE TURKEY

MEDITERRANEAN SEA

CYPRUS

SYRIA

LEBANON

ISRAEL JORDAN

IRAQ

IRAN

AFGHANISTAN

PAKISTAN

TALMESSI—Copper mined about 6000 B.C.

Tr-J

Pb Zn Fe

Zagros Thrust

Sediments over 10,000 m thick

Tigris R. Euphrates R.

Gas

Oil

Abadan

KUWAIT

Shiraz

Tehran Salt Potash

Coal Pb Cu Tr Cu Cr

Tripoli

Alexandria

Cairo

Quattara Depression 133 m below sea level

EGYPT

14-10 m.y.

Sediments 8 m.y. age, 5300 m thick

BAHRAIN QATAR

UNITED ARAB EMIRATES

Gas Gas

PERSIAN GULF

The area between here and the Zagros Thrust contains 65% of known oil reserves

Ophiolites—ocean floor material thrust onto continent about 200 m.y. ago

GULF OF OMAN

Muscat

Murray Ridge

Tropic of Cancer

LIBYA

"Hot spot" volcanic activity 50-0 m.y.

Phosphate in beds 65 m.y.

Fe Cu,Zn

Au

Ni

Au Fe

Fe,Cu,Zn mineralization being formed today

(650-600 m.y.)

650 m.y.

650 m.y.

14-10 m.y.

Au

Riyadh

SAUDI ARABIA

Rub Al Khali Trench Sediments over 9000 m thick

OMAN

Tibesti "hot spot" periodic volcanic activity 50-0 m.y.

Nubian Arabian Craton Orogenies 1000 and 600-400 m.y.

Jidda Fe Mecca

Nubian Arabian Craton Orogenies 1000 and 600-400 m.y.

800-650 m.y.

(1100-800 m.y.)

L. Chad (inland drainage basin)

CHAD

2700 m.y.

Khartoum

SUDAN

Abestos Cr

Au

Pb

NORTH YEMEN

SOUTH YEMEN

Zircon,Ti Sands

N'Djamena

Bomu Complex 3500 m.y.

West Nile Complex 3400-3000 m.y.

65-30 m.y.

Aden GULF OF ADEN

Socotra I.

INDIAN OCEAN

Diamonds

Au

Bangui

CENTRAL AFRICAN REPUBLIC

CAMEROON

Cu

ZAIRE

Fe

CONGO

Congo Basin

KILEMBE Cu

UGANDA KENYA

L. Victoria

ETHIOPIA

Addis Ababa

Fe

Ni

Au

DJIBOUTI

Rift began in Miocene; subsequent uplift continued to Recent time.

Evidence of hominid life dating from 5.5 to 0.5 m.y. in Rift Valley locations from 10°N to 3°S

SOMALIA

Mogadiscio

Carlsberg Ridge

Equator

Red Sea Rift Parting commenced 30 m.y. ago

2,850 to 1,800 m.y.

This overlaps the last period since the first beds of the Witwatersrand series in the basin-shaped structure at Johannesburg date from 2,750 m.y. at which time the orogeny in the Zimbabwe craton to the north was still active. The Witwatersrand has a greater present production of **gold**, has produced more in the past and has more in reserves than any other individual area of the world. There is also a substantial uranium content in some of the conglomerate "reefs" or beds that make the ore. Beds of approximately similar age were deposited around other nuclei throughout Africa but none have been found to have such a remarkable gold and uranium content. The Ventersdorp (2,600-2,350 m.y.) and the Transvaal (2,350-2,100 m.y.) sedimentary series follow, the latter including beds carrying commercial **manganese** and **iron** and also basic intrusives from which **asbestos** is mined. During this period there were three unusual intrusions in southern Africa, each having important economic relationships:—

2,500 m.y.

—intrusion of the Great Dyke (extending for 500 km across Zimbabwe and carrying some of the high-grade **chromium** deposits of that country.

2,300 m.y. (approx)

—intrusion of the pipe-shaped Palabora complex, 400 km northeast of Johannesburg, the carbonatite core of which carries an important **copper** deposit with by-products of iron, phosphate, vermiculite and uranium.

1,950 m.y. (approx)

—intrusion of the Bushveld Complex into the older Precambrian gneisses north of Johannesburg (see inset map on Southern Africa sheet). The surface area of the Complex is about 320 by 240 km, the marginal part being occupied by basic and ultrabasic rocks dipping in towards the centre, which is granite at surface, making a saucer-shaped body fed from a central pipe. The material is believed to be of mantle origin. The Bushveld Complex carries, in the ultrabasic layers, the largest world reserves of **chromium** (but not the best grade) and the largest reserves of **platinum metals** accompanied by some **nickel** and other base metals. **Tin** deposits occur in the upper, granitic layer outcrop in the centre of the Complex.

In the period ending about 1,800 m.y. a major orogeny affected the older cratons and the sediments that had accumulated around them in many parts of Africa. In some cratons, for example those of Maltahohe and Angola on the west side of southern Africa, the evidence of earlier orogenies is almost completely obliterated by the alteration and granitic intrusion of this period. Important base metal deposits associated with sedimentary and volcanic rocks affected by this orogeny include the O'Okiep **copper-gold** and the Tsumeb **lead-copper-zinc** deposit.

1,800 to 550 m.y.

On the eroded surface of the 1,800 m.y. orogeny, and on older basements, sediments were deposited at intervals, interrupted at times by volcanic activity. These are preserved over large areas in central Africa and smaller and more isolated occurrences in northern and southern Africa. Economically, this period is marked by the very extensive bedded **copper** and **cobalt** deposits of Zambia and Zaire and the more sporadic deposits of **uranium** and of **lead-zinc** in the same general series. The **copper-cobalt**-bearing beds (called the "Roan Series" in Zambia) were subsequently folded rather gently. In the northeast part of Africa and adjoining Arabia, however, and also in the southwest, there was a major period of folding, alteration and intrusion dating about 1,000 m.y. which affected previous cratonic assemblages and their surrounding belts of lavas and sediments. Associated with these are some **gold** and base metal ore deposits. Following this there was additional deposition of sediments and lavas ending in the last Precambrian event, which actually extended over into the Palaeozoic since datings range from 600 to 400 m.y. This is known as the "Pan-African Event" and evidence of its intense alteration and intrusion is seen in Saudi Arabia, Egypt and Sudan.

In the same period, dating about 550 m.y., an orogeny that was termed the "Damaran" but is now included under the Pan-African Event, affected western Africa. A quite sharp break, trending north from the Ghana coast at Longitude zero, separates granitic and other rocks dating about 550 m.y. in the area to the east (extending into Nigeria and Cameroon) from that to the west where datings average 2,000 m.y. It is interesting that in the part of the northeastern coast of South America that would fit, prior to continental parting, into the embayment of western Africa, a similar sharp break occurs at Longitude 44°W separating datings of about 2,200 m.y. to the west from those of 650-450 m.y. to the east. This gives yet further support to the concept of plate tectonics.

The Pan-African orogeny of about 550 m.y. also affected part of Namibia. The low-grade but very large **uranium** deposit at Rössing was probably formed at this time.

During the early stages of this period, the late Precambrian orogeny referred to above in northeast and southwest Africa was still active. The central and extreme southern parts of Africa, however, were invaded by seas which, with some withdrawals and advances, deposited sediments during much of this period. There were quite drastic climatic changes including an ice age about 450 m.y. in what is now part of the Sahara Desert. The sediments were accumulated and preserved mainly in two Saharan basins and four basins extending in a rough line down the west centre of the southern lobe of Africa. With the exception of the Cape area at the southern tip of Africa, and the younger Atlas Mountains along the northwesterly edge of the continent, these sediments have remained relatively undisturbed. This was not a period of important mineral deposition but some of the **lead-zinc** deposits in marine limestones later affected by the Atlas Mountain orogeny were probably formed during this period. A relatively small proportion of the **oil** and **gas** occurring in northern Africa is in beds of this period.

350 to 100 m.y.

In northern Africa, there was no basic change from the conditions of the previous period and seas continued to advance and retreat over a large part of the present Sahara area.

In the southern part of the continent, a series of mainly non-marine sediments and volcanics known as the "Karroo Series" contains beds extending in age from 280 to 190 m.y., and is of particular interest since the corresponding series is found in parts of India, South America, Australia and Antarctica, and the comparison constituted one of the important pieces of evidence that led to the concepts of continental drift and plate tectonics. The deposits over this 100 m.y. period in southern Africa occupy now about 500,000 km^2 in the Karroo basin plus large areas (partly covered by younger sediments) in the Botswana (or Kalahari) and Congo basins and originally must have covered a still larger area. Roughly coincident with this deposition, and possibly instigating the Karroo basin, was the development of a system of rift faulting in a general north to northeasterly direction down the east side of the continent. This was a precursor of the Great Rift Valley of eastern Africa.

The Karroo period involved three successive events.

(1) An ice age, which occurred about 280 m.y. in Africa and at slightly earlier or later times in South America, Australia and India (which were still attached to Africa), left tillites, and other deposits typical of glacial events, at the base of the Karroo.

(2) This was followed by the deposition of a thick series of sediments, over 10,000 metres in places, which are dominantly continental, i.e. deposited in inland waters, as indicated by fossils and by dinosaur footprints. There was some relatively minor and local volcanic activity. At several separate periods swampy conditions and abundant vegetation afforded conditions suitable for the formation of **coal** seams which are mined today in several locations in South Africa, Zimbabwe and some adjoining countries in the southern part of the continent. Occurrences of **uranium** have been found at widespread locations in South Africa in sediments of several ages between and below the coal measures. To date (1980) none have reached production but active exploration is in progress.

(3) Towards the end of Karroo times, about 200 m.y., there was a great outpouring of basaltic lavas of the "plateau" type. This event may have been the first indication of the parting of this part of Africa from adjoining parts of Gondwana. These lavas originally extended over a huge area, probably over 350,000 km^2, and although the greater part has been removed by erosion they still cover over 30,000 km^2 in Lesotho and the southerly adjoining part of South Africa, and a further 20,000 km^2 on the eastern and northern borders of the Republic.

The sediments in the south Cape region, mainly of Devonian (395-345 m.y.) but involving some Karroo beds, underwent folding about 200 m.y. producing fold axes that are orientated in an eastwest direction along the south coast swinging northerly in the vicinity of Capetown. This is the only folding in Africa south of the equator after Precambrian times.

The major event in the latter part of this time interval was the splitting apart of the continents that made up the Gondwana portion of the Pangaea supercontinent. These partitions probably started with rifts (similar to those seen in the Great Rift Valley of Africa today) up which rose lavas, first forming the plateau basalts referred to above and later forming ocean floor as the new continents drifted apart. The dating appears to have been about 200 m.y. for the first parting of Africa from South America and Australia and probably somewhat later from India. Later still, Madagascar split off from the east side south of Somalia.

An event in Cretaceous times (135 to 65 m.y.) of geological and economic significance in southern and west central Africa was the intrusion of clusters of pipes and dykes of kimberlite, an ultrabasic rock usually carrying inclusions of a variety of rocks through which the molten material has passed in its passage from a very deep source within the Earth's mantle. Kimberlites are the original source rock of gem and industrial diamonds but only a relatively minor proportion of the pipes or dykes carry the mineral in commercial proportions. Some pipes are believed to have been intruded about 1,000 m.y. ago but the period about 100 m.y. (after the end of Karroo times) seems to have been one of particularly widespread kimberlite activity. Erosion subsequent to the consolidation of the kimberlites and their exposure at surface has resulted in diamonds being released and, since the mineral is extremely resistant to wear or chemical attack, being concentrated in the gravels or rivers or coastal beaches in quantities to make a second important source for commercial recovery. Such gravels may be of any age subsequent to the exposure at surface of the kimberlites but most of those worked commercially are quite young, less than one million years. Examples are gravels in the valley of the Orange River, flowing westerly from a region of kimberlite pipes to the west coast, and beach gravels on the west coast of South Africa and Namibia and also in the seas off this coast.

100 m.y. to Present

During this period, large areas of Africa received sedimentary deposition partly from marine invasions in the north and partly from fresh water bodies. In most parts of Africa and Arabia these have remained more or less undisturbed by folding with exceptions in two areas:—

(1) Oman, on the extreme east side of the Arabian Peninsula, was involved about 90 m.y. in a chain of folding seen mainly in Pakistan, Iran and Turkey. One of the effects in Oman was the thrusting onto the edge of the continent of ultrabasic material from the ocean floor—ophiolites—which are also seen on the opposite side of the Gulf which now separates Arabia from Eurasia. **Copper** and **pyrite** associated with these rocks (as in Cyprus) were worked in the early stages of Man's use of metals.

(2) In the northern coast of the continent, between Tunisia and western Morocco, a belt about 2,500 km long and 350 km wide is occupied by highly folded sediments resulting from the Atlas orogeny. This, like the one in Oman, is an expression of more widespread folding in southern Eurasia (in this case dating from about 90 to 50 m.y.) and resulted from the convergence of the African and Eurasian Plates.

From the economic standpoint deposits of **lead, zinc** and **manganese,** with minor amounts of other metals, occur within the Atlas fold belt but may have been formed prior to the folding. The most important deposits in this part of Africa are non-metallic, especially:—

(1) **Phosphate** deposits in Morocco, Algeria and the former Spanish Sahara; those in Morocco constitute the largest world reserves. These are in beds of about 65 m.y.

(2) **Oil** and **gas** accumulations in sediments dating from about 45 m.y. to relatively recent times are found near the coast in nearly all the countries bordering the Mediterranean and extend a considerable way beneath the Sahara Desert in Algeria and Libya. The petroleum accumulations on the delta of the Niger River, both on and offshore, are in sediments of quite recent geological age, as are those in two areas further south on the west coast.

By far the greatest accumulation of petroleum in the world is in and under the shores of the Gulf that separates Saudi Arabia from Iran. An area approximately 1,400 by 600 km contains about 55% of the world's known oil reserves and accounts for 37% of current production. Additional concentrations are in southern Oman on the southeastern edge of the Arabian Peninsula. The greater proportion of these oil reserves is in sediments dating between 37 m.y. and 20 m.y. and the same series contains considerable thicknesses of salt and other evaporite beds. Such chemical deposits commonly form when an arm of the sea becomes cut off from direct contact with the ocean and, by evaporation, becomes over-saturated successively in different salts and precipitates them to form beds of rock salt, gypsum, potash, etc. If the connection with the outside ocean is re-opened and then closed again the process is repeated. This condition seems to have occurred to a remarkable degree in the area now surrounding the Arabian (Persian) Gulf. Such conditions also favour the accumulation and burial of the forms of pelagic life that are the source of petroleum. In addition to sharing the same sedimentary conditions that are most favourable for the generation of petroleum, saline deposits play an important part in being plastic under pressure but impervious to oil migration so that they make ideal traps for oil and gas accumulation.

In northeastern Africa, the last geological event of fundamental importance was the splitting off of the Arabian Peninsula from the main body of the African continent. This occurred along a fracture that evidently first opened about 30 m.y. and resulted in the initiation of the Red Sea. There was a pause in the parting process and it was renewed about 10 m.y. and is still continuing. The Red

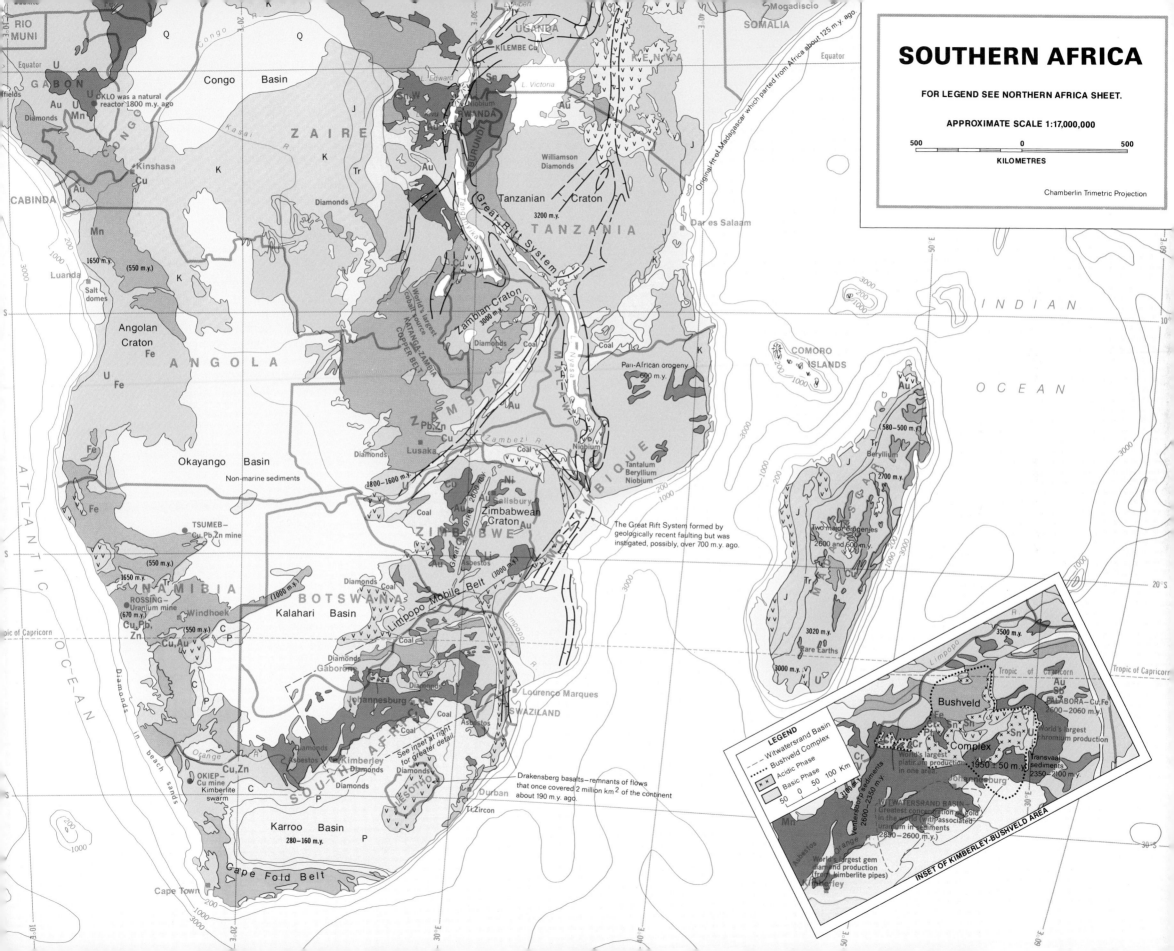

SOUTHERN AFRICA

FOR LEGEND SEE NORTHERN AFRICA SHEET.

APPROXIMATE SCALE 1:17,000,000

500   0   500

KILOMETRES

Chamberlin Trimetric Projection

RIO MUNI

GABON

Oilfields

Au U

OKLO was a natural reactor 1800 m.y. ago

Diamonds

Au

Mn

CABINDA

Au

Kinshasa

Cu

CONGO

Congo   Basin

ZAIRE

Q

Q

Congo

Kasai R.

1650 m.y.   (550 m.y.)

Luanda

Salt domes

Angolan Craton   Fe

ANGOLA

U   Fe

Fe

Okayango   Basin

Non-marine sediments

TSUMEB – Cu Pb Zn mine

Fe

(550 m.y.)

1650 m.y.

NAMIBIA

ROSSING – Uranium mine (670 m.y.)

Windhoek

Cu,Pb, Zn

Cu,Au

(550 m.y.)   C   P

Diamonds in beach sands

Orange R.

Cu,Zn

OKIEP – Cu mine Kimberlite swarm

C

P

ATLANTIC   OCEAN

Cape Town

UGANDA

KILEMBE Cu

L. Edward

Sn W

RWANDA

BURUNDI

KENYA

Williamson Diamonds

Tanzanian   Craton

3200 m.y.

L. Victoria

Au

Niobium

Dar es Salaam

TANZANIA

Great Rift System

World's largest cobalt source KATANGA-ZAMBIA COPPER BELT

Zambian Craton 3000 m.y.

Cu

Diamonds

Z Pb Zn

Lusaka

Cu

Diamonds

ZAMBIA

L. Tanganyika

L. Nyasa

MALAWI

Coal

Coal

Pan-African orogeny 600 m.y.

1800-1600 m.y.

Zambezi R.

Coal

Niobium

Tantalum Beryllium Niobium

MOZAMBIQUE

Co

Au

2600 m.y.   Au   Salisbury

Ni

Coal

Au

Great Dike

Zimbabwean Craton

ZIMBABWE   Au

Li

Asbestos

(3000 m.y.)

Au

Limpopo   Mobile   Belt

(3000 m.y.)

The Great Rift System formed by geologically recent faulting but was instigated, possibly, over 700 m.y. ago.

Original Rift of Madagascar which parted from Africa about 125 m.y. ago

Equator

SOMALIA

Mogadiscio

INDIAN   OCEAN

COMORO ISLANDS

580-500 m.y.

Tr   Beryllium

Au

2700 m.y.

Two major orogenies 2600 and 600 m.y.

Cu

MADAGASCAR

Tr

3020 m.y.

Rare Earths

3000 m.y.

U

BOTSWANA

Kalahari   Basin

Coal

Diamonds

Gaborone

Diamonds

(1000 m.y.)

Coal

Au

Cu

Coal

Johannesburg

Diamonds

Asbestos

Kimberley

Diamonds

OKIEP – Cu mine Kimberlite swarm

SOUTH AFRICA

See inset at right for greater detail

Asbestos

Coal

SWAZILAND

LESOTHO

Durban

Tr, Zircon

Drakensberg basalts—remnants of flows that once covered 2 million km² of the continent about 190 m.y. ago.

Karroo   Basin

280-160 m.y.

Cape Fold Belt

Lourenço Marques

Limpopo R.

INSET OF KIMBERLEY-BUSHVELD AREA

3500 m.y.

Limpopo R.

Tropic of Capricorn

Bushveld

Fe

Cr

Pt

Cr

Sn

Sn

Complex

World's largest platinum production in one area

Au   Sn

PALABORA – Cu, Fe 2600 – 2060 m.y.

World's largest chromium production

1950 – 50 m.y.

Transvaal sediments 2350 – 2100 m.y.

Ventersdorp sediments 2600 – 2350 m.y.

WITWATERSRAND BASIN Greatest concentration of gold in the world (with associated uranium in sediments 2850 – 2600 m.y.)

Johannesburg

Asbestos

Mn

World's largest gem diamond production (from kimberlite pipes)

Kimberley

LEGEND

Witwatersrand Basin

Bushveld Complex

Acidic Phase

Basic Phase

50   0   50   100 Km

Sea, and the Gulf of Aden into which it turns at a sharp angle, thus constitute an ocean in infancy as shown by the dating of lava flows that have come up along the fractures (there are two main parallel Red Sea fractures) and by the thickness and age of sedimentary covering over the volcanic material on each side of the fractures. Of great interest to the economic geologist was the discovery in 1965 of an ore deposit in the process of formation at a point on the rift in the Red Sea. **Iron** and **copper,** with other metals in smaller proportions, are being precipitated at a place where the water is hotter and more saline than the surrounding average.

Of still more recent origin than the Red Sea is the Great Rift Valley extending from the south end of the Red Sea southerly to about Latitude $10°$S where it splits into several branches. As has been mentioned earlier, rift faulting had already occurred about 200 m.y. along the same general lines in southern Africa but had remained inactive since then until about 30 m.y. The main part of the recent rift pattern is bordered by major north-trending faults and marked by

floods of lava that have come up during the past 30 m.y. with some centres still periodically active. If the Red Sea may be described as an ocean in infancy, the African Rift Valley should be classed as an ocean in gestation. There is little doubt that this rift system is the beginning of a parting process that will result, millions of years hence, in that part of Africa east of the Rift Valley becoming a separate land mass.

It is perhaps a curious coincidence that the Great Rift Valley, signifying the youngest major geological event in Africa, and perhaps in the world, should include the sites of the earliest evidence of the human race. Skeletons and crude artifacts of Australopithecus Africanus, regarded as the oldest hominid, have been found at several places between the Olduvai Gorge in Tanzania and Omo in Ethiopia within the general confines of the Rift Valley. These discoveries can be dated, in some places, by reason of lying above or below volcanic tuff beds that are suitable for radiometric dating. These show that some of the artifacts and fossil bones in northeastern Kenya go back to over 3 m.y. before the present and further south less firm evidence suggests dates as far back as 5 m.y.

THE USSR SHEET

A glance at the map covering the territory enclosed by the boundaries of the USSR shows at once that it is unevenly divided into a smaller European part and a larger Asiatic one. The divider is the Ural Mountain chain, which is indicated by the north-trending band of roughly parallel but complex geological units along Longitude 60°E. In each of the western and eastern parts, the broad geological structure is built around a Precambrian nucleus. In the west, this nucleus is the Baltic Shield (extending into Finland) which is the exposed part of the "Russian Platform" on which younger, undisturbed sediments have been deposited. A large area of Precambrian is exposed to the southwest in the Ukraine which is separate from the main Russian Platform.

East of the Urals the circular area of exposed Precambrian about Longitude 110°E forms the nucleus of the "Siberian Platform" around which roughly concentric bands of successively younger sediments lie undisturbed on the Precambrian basement.

The southern and eastern parts of the area covered by this map show the complex pattern resulting from successive waves of folding and intrusion in later stages of the geological history. These events affected younger sediments and volcanics as well as the underlying Palaeozoic and Precambrian, a large area of the latter being exposed between Lake Baykal and the Pacific Coast.

The explanation of the almost unique occurrence of a mountain chain (the Urals) in the middle of a continent is believed by most western earth scientists (but by no means the majority of Soviet ones) to have resulted from the collision, starting about 400 m.y. ago, and eventual attachment of two tectonic plates originally widely separated by ocean floor. These two plates have been named respectively the European and Asiatic Plates, the joining of which formed the present Eurasian Plate which includes Asia, Europe and the eastern Atlantic. It has since remained as a single tectonic unit.

A vast area of roughly east-west folding, with accompanying volcanic and intrusive activity, occupies southern USSR. The Pacific side is characterised by dominantly northerly, but irregular, trends of folding. These features have resulted from a very complex interweaving of tectonic events. Generally, however, the successive orogenies seem to have been caused by the spasmodic convergence of other continent-bearing plates from the south side onto the southern edge of the Eurasian Plate and the subduction of the oceanic crust of the Pacific Plate under its eastern edge.

The USSR, occupying nearly one-sixth of the land surface of the Earth, is well supplied with mineral resources including fossil fuels and most of the metals required by a balanced industrial economy. The known reserves of coal and iron ores are particularly abundant, extending over wide geographic limits as well as over greater geological time intervals than in most other parts of the world. Although there are, to date, no known individual orebodies of non-ferrous base metals of worldwide fame, enough such deposits exist to give self-sufficiency in most metals and elements. Tin is an exception and has been mainly imported and lead was sufficiently scarce for USSR to be a net importer in 1976. Even in the case of tin, however, discoveries have been made over the past 20 years in granites in the Pacific region. In the chronological summary that follows mineral deposits are described relative to geological events.

Pre-2,500 m.y.

The Archaean history of the USSR is probably less completely known than that of the Americas, Africa or Australia because a much smaller proportion is exposed. Broadly, however, the same succession of events seems to have taken place on both the European and Asian Plates which, as previously mentioned, are believed by many earth scientists to have been separated by ocean at this time. The earliest event so far recorded is a period of dominant volcanic activity, with subordinate sedimentation, affected by probably several phases of folding, intrusion and metamorphism culminating about 2,500 m.y. ago. Subsequent erosion has resulted in belts of predominantly basic lavas and volcanic sediments separated by broader areas of gneisses and granites.

The mineral deposits associated with this early geological history include **iron** formations, and some minor **nickel** deposits, in the Kola Peninsula and in the circular Precambrian nucleus in northern Siberia. Unlike those in most other parts of the world very few of the lode **gold** deposits in Russia are in Precambrian rocks, most being in volcanic belts, and related intrusions, of Palaeozoic and later age.

2,500 to 1,700 m.y.

In this period, referred to by Soviet geologists as "Lower Proterozoic", erosion of the rocks affected by the previous orogeny was followed by the deposition of locally thick series of sediments and volcanics on the old basement and on the flanks of the original platform. These units may have been affected by several orogenies but certainly by a major one culminating about 1,700 m.y., which extended over wide parts of both the European and Asian Plates and greatly increased the areas of the Russian and Siberian Platforms.

Mineral deposits formed during this period include the very important bedded **iron** deposits of Krivoy Rog, near the Black Sea, where iron has been worked for many centuries, and at Belgorod and Kursk, 500 km south of Moscow where one of the largest magnetic anomalies in the world indicates a huge potential of ore. **Iron** ores of this age also are found in the Precambrian area of Angara in southern Siberia. The **nickel** deposits of Pechanga near the Finnish border are dated about 1,770 m.y.

1,700 to 570 m.y.

This time interval "Upper Proterozoic" in Soviet terminology (i.e. the last division of the Precambrian), is represented to some extent in the Kola Peninsula, the Ukraine Precambrian area and in the two main Siberian areas. The rocks consist mainly of sediments with some volcanics. These have been little disturbed where they lie on the older Precambrian platforms. An orogeny, in most places much milder than those referred to in older Precambrian, occurred between 1,000 m.y. and the end of this period. This event has been called the "Baykalian" after Lake Baykal, which is on the western margin of the Precambrian area, where the orogeny is most pronounced.

This was not a period of great mineral deposition but some Kola Peninsula **nickel** deposits are probably of this age.

570 to 370 m.y.

The European and Asian Plates are believed to have remained separated during this period (Early Palaeozoic) by a deepening trench, but were converging near its close. In both plates the Precambrian platforms seem to have been sufficiently reduced, by erosion and subsidence, to allow the flooding of seas over extensive areas. The relatively undisturbed sediments deposited by these inland seas contain the largest **salt** deposits in USSR, in the Cambrian beds on

the Angara River. **Phosphate**-bearing sediments of equivalent age occur in the mid-Asian region.

Around 500 m.y., folding occurred in the deep troughs containing rapidly accumulating sediments, such as the trough that later became the Ural Mountains and others south of the Siberian Platform. Folding events continued in surges broken by quiescent intervals, but in general the folding, intrusion and metamorphism reached a peak between 420 and 400 m.y. in the western belt; between 370 and 280 m.y. in the middle and between 320 and 240 m.y. in the east. This orogeny has been called the "Caledonian" after the period of folding in northern Britain and Scandinavia of this age, although there was probably little direct relationship between the two. This folding formed the Ural Mountains trending north and probably also the northeasterly-trending ranges in Novaya Zemlya and the Taymyr Peninsula.

Deposition of metallic minerals in this period is illustrated by the **lead** and **zinc** mineralisation in carbonate beds on the Russian and Siberian Platforms. Important **manganese** deposits occur in beds of about 470 m.y. and 370 m.y. in the Angara region of Siberia. In the area south of the Aral Sea lies the Muruntau Mine which in 1978, was the largest individual gold producer in the world having an output, from an openpit operation, of 80 tonnes against the largest Witwatersrand individual mine (Vaal Reefs) with 72 tonnes. It was probably emplaced in its present form about 450 m.y. but may have existed originally in Precambrian formations.

In the Ural Mountain chain there is a great variety of metallic deposits including **iron, manganese, copper** and **nickel.** Some of these were originally deposited in beds prior to the Ural folding which marked only the first stage in the formation of this chain. The most important sources of **platinum** ores, associated with nickel, occur in ultrabasic rocks that are believed to be ocean floor material (ophiolites) thrust onto the edges of continents in the course of Ural disturbances. The Russian **chrome** deposits, of world significance, are also associated with ultrabasic rocks of the same probable origin and age. Important production of **asbestos** is obtained from similar ultrabasic rocks.

Some, but not the majority, of the **oil** and **gas** reserves of USSR are contained in sediments of this general period from about 500 to 370 m.y.

370 to 225 m.y.

This period, Late Palaeozoic, saw the dying stages of the orogeny that produced the Ural Mountains and the attachment of the European and Asian Plates. The extensive deposition of freshwater sediments on both the European and Siberian Platforms, as well as in southern Siberia, is indicative of a general uplift and recession of seas at the beginning of this period. In isolated embayments of the sea, **salt** deposits were formed such as those now mined from beds and salt domes on the northern margin of the Siberian Platform. **Manganese** beds of Devonian age (about 370 m.y.) are mined in Kazakhstan where sediments and volcanics of the period 300-390 m.y. are exposed. This region is also the most important source of **copper** in the USSR; the deposits occurring both in sandstones of about 300 m.y., and also in the region of Lake Balkhash in the form of porphyry copper.

The rapid development of spore-bearing tree-fern flora and suitable climatic conditions for swamps introduced the extensive geological period from about 340 to 130 m.y. during which **coal** was formed at widely separated parts of USSR. Some of the oldest fields are north of Moscow but there are large reserves on both flanks of the Urals, in central Siberia on the west side of the Siberian Traps (plateau basalts) and in the Pacific region.

The folding in southern USSR that culminated about 275 m.y. ago has been called "Hercynian" after folding of similar age in western Europe of which it is probably a continuation. There may have been no sharp discontinuity between the folding of the Urals and the Hercynian folding even though the former has a northerly orientation and the latter roughly east-west, although with sinuous variations. Associated with the Hercynian orogeny, or at least affected by it, are important **lead-zinc-silver** deposits in the Angara region in south-central Siberia. A large disseminated **copper** deposit currently under development near Udokan is probably of the same general age.

Oil and **gas** are found in the marine sediments (and in some places these have migrated into terrestrial sandstones) of this period, especially west of the Ural Mountains.

A major event of geological significance, overlapping the Hercynian folding and intrusion, was the outpouring of basic lavas in east-central Siberia about 200 m.y. These, known as the "Siberian Traps", now occupy an area of close to 750,000 km² and probably originally covered much more. They consist of bedded tuffs near the base, and basalt flows at the top. Related intrusive sills and dykes are interlayered within the volcanic trap rocks surrounding older formations. Unlike most major plateau basalt areas in other parts of the world, they do not appear to be connected with rift faulting that initiates continental parting. The explanation of some Soviet geologists is that the magma has been forced up by the depression of adjoining parts of the Siberian Platform. A substantial tonnage of **magnetite iron** ore occurs in several deposits in or near the basaltic traps. An important source of **platinum** metals (of which USSR is the second largest world producer) is a by-product from the working of **nickel-copper** sulphides which occur in intrusive basic rocks dating about 200 m.y. near Norilsk at the northern edge of the main mass of the "Siberian Traps".

The **diamond** mining industry is comparatively new in USSR but has made considerable progress. The most productive district is in and near the circular Precambrian area east of the Siberian Traps where a cluster of kimberlite pipes and dykes were injected about 290 to 250 m.y. Kimberlites, the primary source of diamonds, originate in the mantle at depths of 30 km or more and are most productive where they penetrate early Precambrian cratons.

225 to 26 m.y.

This time interval, covering the Mesozoic and early Tertiary, follows the decline of the Hercynian folding and intrusion which had affected mainly the southern third of the USSR and coincided with the last phase of folding in the Urals. In the northern two-thirds of the USSR this was a period of quiet, fresh-water deposition of sediments, and the accumulation of plant matter now observed as **coal** seams, particularly in Siberia. The earliest of these sediments followed, or were contemporaneous with, the extensive basalt flows described above and extend from the west border of the USSR to the Pacific coast.

A new period of folding and intrusion, which culminated in great outpourings of lavas, affected especially the Pacific region of the USSR and to some extent also the south and extreme west. This orogeny, known as the Cimmerian, extended from approximately 200 to 120 m.y. The folding was not as intense as some of the earlier orogenies but was characterised by numerous granitic intrusions. The latter stages of this event involved great volcanic activity in the late Cretaceous—from about 100 to 65 m.y.

Sedimentary **iron** of economic significance in beds between 150 and 140 m.y. old is mined in western USSR. Iron ore of about the same age is also of importance near Tomsk in the Western Siberian Lowlands, and in the area between the Caspian Sea and the Turkish border. The large Nicopol **manganese** deposit is in Oligocene sediments (about 35 m.y.) not far from the Krivoy Rog iron deposits in southern Ukraine. **Bauxite** occurs in several sedimentary horizons between 100 and 50 m.y.

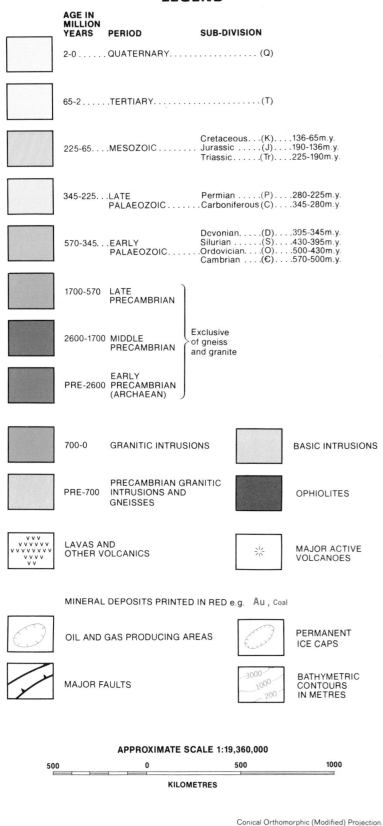

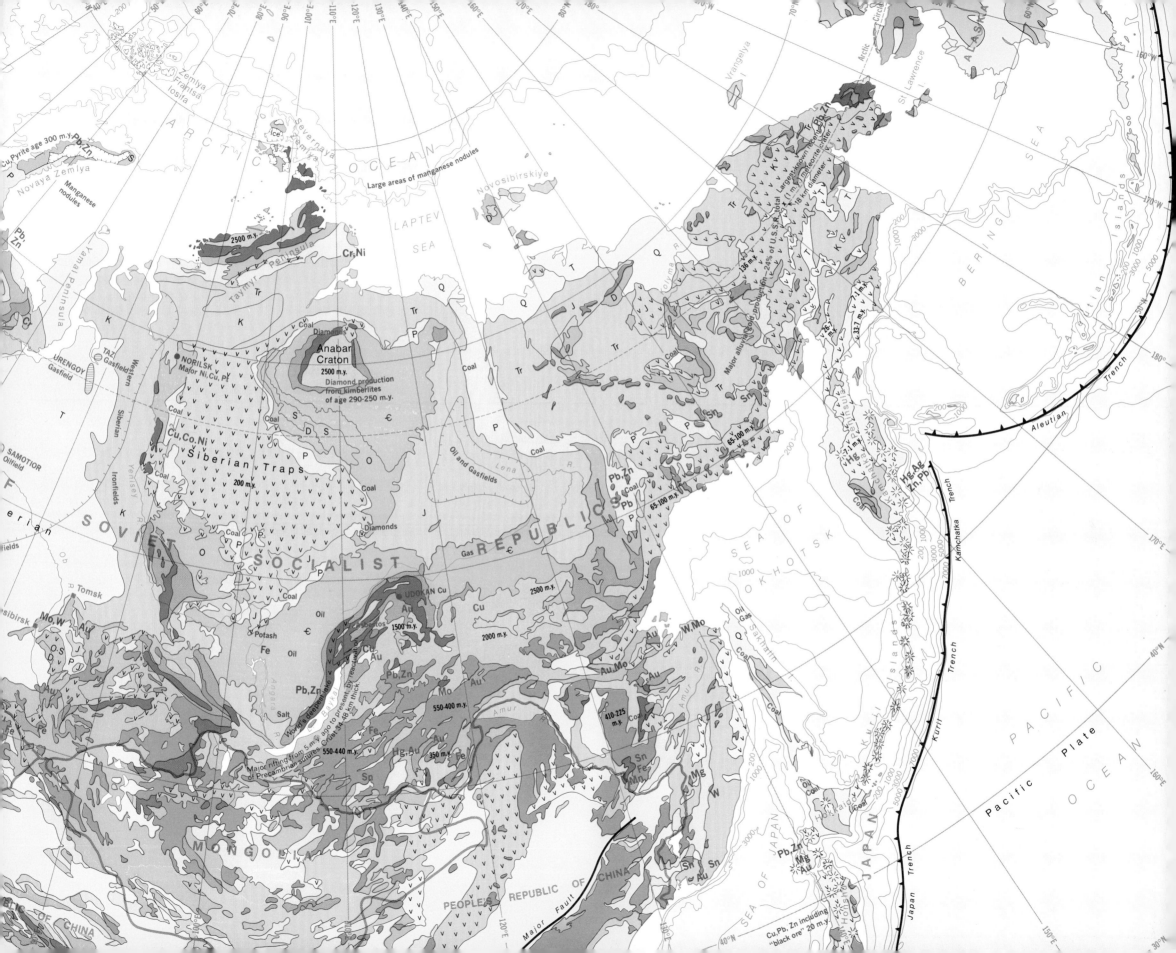

Tin, one of the few metals in which USSR is not yet believed to be self-sufficient, occurs in numerous small deposits in the form of veins in granite stocks of Cimmerian (200 to 120 m.y.) age in the Pacific region. The extent of the reserves is not known but the area is receiving active exploration for additional occurrences.

Most of the known reserves of **oil** and **gas** in the USSR are in marine sediments deposited during this time interval. The most extensive basin is in the west Siberian Plain, east of the Ural Mountains, but there is also major production from the vicinity of Kharkov, west of the Urals, and between the Aral Sea and the border with Iran and Afghanistan.

26 m.y. to Present

Sedimentation continued in the mid-Asiatic region and in the general area surrounding the Aral Sea. Incursions of the sea deposited sediments on the Kamchatka Peninsula on the Pacific coast. The volcanic activity of the previous period on the Pacific coastal area was repeated on this peninsula and north of it as may be seen on the map.

The major event of this period was the onset of the Alpine orogeny which started about 55 m.y. but did not reach its maximum mountain building activity until the Pliocene (7 m.y.). Although the most spectacular results of this orogeny are outside the boundaries of the USSR, it had a marked effect along the 800 km wide belt north of the south boundary from the Pacific coast west to Kazakhstan. Since this belt had already been affected by at least three orogenies, it is difficult to define the Alpine structures. Certainly, it is strongest next to the south border and weakens going north.

The Pacific region of the USSR was affected to some extent by the Alpine orogeny but more by intrusion and volcanic activity than by folding.

Over 55% of the **gold** produced in 1976 in the USSR was derived from placer deposits formed in the last period of geological history although the metal was derived from the weathering and erosion of older rocks or deposits. The largest area of placer gold workings, responsible for about 45% of the 1976 placer production and 24% of the total gold output, is in the Severovostok area in the extreme northeast of Siberia, which was not swept by continental ice caps in the last Ice Age.

Another deposit of importance of post-26 m.y. is that of sedimentary **iron** ore at Kerch on the north coast of the Black Sea.

Some of the **oil** produced from the Central Siberian Plain is from sediments of less than 26 m.y. age.

One of the most recent geological events of significance within the USSR territory shows the most visible results in the vicinity of Lake Baykal (the deepest inland body of water in the world) which forms the centre of a linear system of rifting that extends 3,000 km from northern Mongolia through eastern Siberia to southern Yakutia. The current episode of rifting began about 5 m.y., but was probably controlled in position and direction by faults that date back to Precambrian times. A very rapid depression has occurred resulting in between 5,000 and 7,000 metres thickness of sediments. In addition, volcanic activity began prior to the rifting some 20 m.y. ago and continues to some extent today, as expressed by hot springs and high lake-bottom temperatures. The tectonic activity in the area is obvious when it is realised that there are on an average 10 earthquakes a day, some of small magnitude, but about 20 a year have a magnitude of greater than 4.5 on the Richter scale.

In the most recent period of geological history (beginning about 600,000 years ago), the northwestern part of the area covered by this map sheet was swept by ice sheets of the Pleistocene Ice Age. The southern limit of the continental glaciation extended in two prongs reaching within 650 km of the Black Sea, then in a northerly line west of the Volga River to about Latitude 60°N, along this to the Yenisey River and then in a series of northerly and easterly lines to reach the Arctic Ocean at Longitude 118°E. Thus the northeasterly part of Siberia was not glaciated, other than in local spots of high altitude, which has resulted in the preservation of the important alluvial gold referred to earlier.

The still-existing remnants of the ice sheet in the northwest, in the form of small ice-caps on northern islands and valley glaciers, have yielded a number of excellently preserved mammoths and other now extinct mammals that lived along the margins of the glaciers. The lack of ice-caps over northeastern Siberia permitted the spread of the human race over this area and subsequent migration over the Bering Straits to North America, probably over 40,000 years ago.

THE SOUTHERN ASIA SHEET

The geological and tectonic complexity of the southeastern part of the Eurasian continent is seen on this map sheet. This complex pattern results in large part from an unusually violent collision between two continental bodies.

It may be seen that Precambrian rocks, which form the foundation of all continents, are exposed in three areas within this sheet:

(1) Peninsular India and the island of Sri Lanka (Ceylon).

(2) A group of separate patches in eastern China between Latitudes 26° and 34°N and Longitudes 108° and 118°E.

(3) Strips of Precambrian following the general trend of Himalayan folding, having been involved in this and previous orogenies and then re-exposed by subsequent erosion.

It may also be observed that uncovered rocks of Palaeozoic (from about 570 to 225 m.y.) occupy a relatively small proportion of the land surface and much less than the area occupied by rocks less than 225 m.y. old

There is now convincing evidence that, prior to 200 m.y. India (together with the island of Sri Lanka) was attached to the southern part of the east side of the African continent forming part of Gondwana, the southern member of the former massing of continents, and thus was separated by 5,000 km of ocean from southern Asia. What is less certain is how much of the area now lying north of the Indian Precambrian shield also formed part of Gondwana. The succession of thrust slabs that form the Himalayan mountains consist mainly of sediments that were probably deposited on the continental shelf north of India. Even the area north of Tibet is thought by some authorities to have been part of the Indian "microcontinent". The junction of two continents that have collided, known as a 'suture", is often difficult to identify since it becomes sealed and distorted by subsequent tectonic disturbances. One quite definite suture can be traced in a curved line following the Indus and Tsengpo valleys by the presence of ocean-bottom material (ophiolite) that has been thrust and folded between the converging plates.

To the east the area from Burma to the Malaysian Peninsula, and everything east of this, was probably always part of Asia proper.

The break-away of the Indian microcontinent from Africa started with rifting that occurred about 190 m.y. ago in a roughly north-south direction coinciding approximately with the present east side of southern Africa. Approximately contemporaneous with this rifting was the great outpouring of basaltic lavas over a large part of southern Africa. Actual parting along the rift zones commenced at about 140 m.y. and allowed the uprising of oceanic lavas (as is now taking place at the mid-Atlantic Ridge) forming a new incipient ocean. As the parting widened, the newly-formed Indian continent moved east and then started its relatively rapid drift to the north.

In the meantime, the southern part of Eurasia had grown by the addition of sediments and volcanics on the southern continental shelf of the original Precambrian basement. It may also have been added to by the arrival, collision and adherence of what is now Iran, Afghanistan and Pakistan west of the Indus. This continental fragment is believed, on the basis of recent evidence, to have been attached to the east or northeast side of Arabia (which at this time was solidly attached to the African continent) and to have broken away some time after 190 m.y. as a microcontinent and to have drifted northeast and collided with Eurasia about 130 m.y., well before the arrival of India.

The distance travelled by the plate carrying the Indian continent from the time of its parting from Africa about 140 m.y. until its collision with Eurasia was approximately 5,000 km. The average speed was considerably more than most measurable plate movements and it reached its maximum, probably about 20 centimetres per year, between 70 and 55 m.y. Folding and thrusting of the sediments and volcanics on the south flank of Asia evidently occurred before the main collision and accounts for the existence of other sutures of increasing age going north. The main collision appears to have occurred about 50 to 40 m.y. but obviously the first contact would have been at projecting points on the north side of the Indian continent or the south side of Eurasia and the completion of the collision would have taken some millions of years. The collision did not stop the northern movement of the Indian continent which continued to push into the southern side of Eurasia, albeit at half its former speed, for a further 2,000 km and is still continuing today.

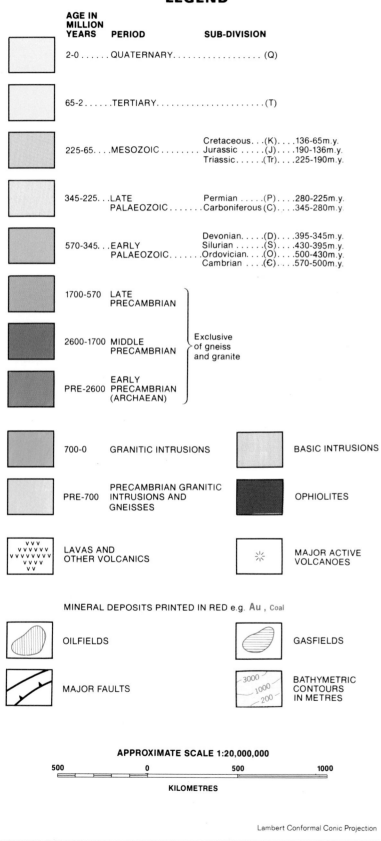

SOUTHERN ASIA

LEGEND

AGE IN MILLION YEARS	PERIOD	SUB-DIVISION
2-0	QUATERNARY	(Q)
65-2	TERTIARY	(T)
225-65	MESOZOIC	Cretaceous...(K)....136-65m.y.
		Jurassic.....(J)....190-136m.y.
		Triassic......(Tr)....225-190m.y.
345-225	LATE PALAEOZOIC	Permian.....(P)....280-225m.y.
		Carboniferous(C)....345-280m.y.
570-345	EARLY PALAEOZOIC	Devonian.....(D)....395-345m.y.
		Silurian......(S)....430-395m.y.
		Ordovician...(O)....500-430m.y.
		Cambrian...(Є)....570-500m.y.
1700-570	LATE PRECAMBRIAN	
2600-1700	MIDDLE PRECAMBRIAN	Exclusive of gneiss and granite
PRE-2600	EARLY PRECAMBRIAN (ARCHAEAN)	

700-0   GRANITIC INTRUSIONS

BASIC INTRUSIONS

PRE-700   PRECAMBRIAN GRANITIC INTRUSIONS AND GNEISSES

OPHIOLITES

LAVAS AND OTHER VOLCANICS

MAJOR ACTIVE VOLCANOES

MINERAL DEPOSITS PRINTED IN RED e.g. Au , Coal

OILFIELDS

GASFIELDS

MAJOR FAULTS

3000 / 1000 / 200   BATHYMETRIC CONTOURS IN METRES

APPROXIMATE SCALE 1:20,000,000

500   0   500   1000

KILOMETRES

Lambert Conformal Conic Projection

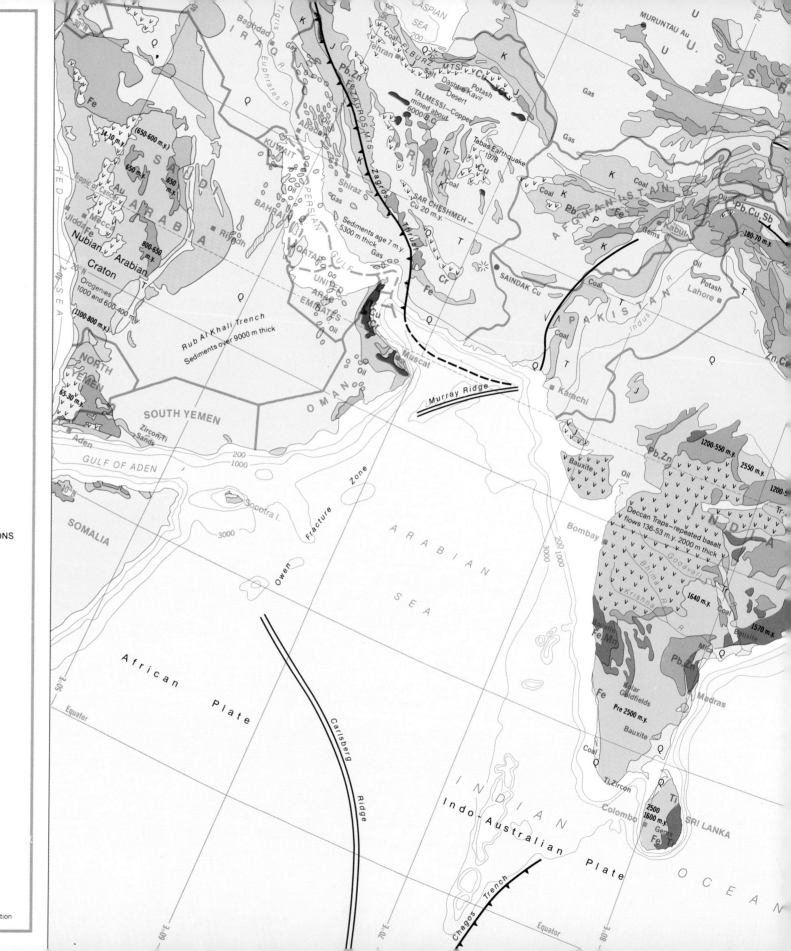

The way the rocks of Eurasia "made room" for this massive invasion has been credited to three different processes:

(1) By the shortening in a north-south direction of the sedimentary and volcanic rocks lying ahead of the advancing ram by means of crumpling and overthrusting. This could only explain a proportion of the 2,000 km shortening to be accounted for.

(2) By subduction of part of the Indian Plate under the Eurasian continent. Although there is little doubt that subduction of the oceanic crust occurred ahead of India, the susceptibility of continental crust to subduction is normally rather limited due to its relative lightness and thickness.

(3) By the rocks ahead of the advancing Indian continent being pushed aside—especially on the east side—by means of lateral movement on faults approximately normal to the direction of the push. The action is somewhat similar to the behaviour of a sheet of ice being broken and pushed aside by the bow of a moving ship.

All three processes probably played a part in making way for this massive invasion but the relative importance of each is still a matter for debate.

The furthest north of the advancing "ram" may be seen on the map to be near the area where the north easterly parts of both Pakistan and Afghanistan approach the Chinese border. West and southwest of this, the folds tend to be "left behind" the advancing ram with resulting counter-clockwise rotation of the fold axes.

From the time of the main collision there was substantial vertical movement as clearly shown by:

(1) Marine sedimentation in several periods subsequent to 75 m.y. proves that large areas north of the present Himalayas were below sea level but are now far above it; e.g. thick limestone of about this age forms the Tibetan plateau which now lies at an elevation of over 5,000 metres.

(2) River valleys that cross the Himalayas from north to south must have flowed from higher ground in Tibet to the Indian Ocean prior to the uplift of this great range.

Since the Indian sub-continent and original Asia were separated by 5,000 km of ocean during geological history prior to 140 m.y., it is logical to deal with the two separately, at least up to the date when India broke away from Africa. For this purpose the term "Original India" is used to include, besides Sri Lanka and peninsular India, a large area to the north of it bounded by a curved line commencing about the mouth of the River Brahmaputra and extending up it north of the Himalayas along the Tsengpo and Indus valleys and then southwest and south just west of the mouth of the Indus River. This area, which has subsequently been reduced in a north-south direction by intense folding probably formed the continental shelf that lay north of the Precambrian triangle of "Original India" and travelled north with it.

As mentioned, Iran, Afghanistan and Pakistan west of the Indus may have constituted a separate addition to the Eurasian mainland but, for present purposes, all the land west, north and east of the curved Indus-Tsengpo suture, which is believed to have formed the main line of continental collision, is referred to as "Original Asia".

ORIGINAL INDIA

Pre-2,500 m.y.

The southern and western part of the Indian Precambrian Shield shows a pattern of relatively narrow belts of volcanics and related sediments ("greenstones") separated by wider zones of granite and gneiss, that is similar to the Archaean parts of the Precambrian of Zimbabwe, Western Australia and Canada. Volcanic activity was evidently frequent and violent.

The mineral deposits that were formed during this period include **gold** in vein-type deposits which have been mined, especially in Mysore, for many years and to considerable depths. In the same greenstone belts **iron** formations are found, originally quite low grade (30% Fe approximately) but which have been, in some places, enriched by the action of surface solutions. The more important iron deposits of India were formed in later Precambrian.

Exposures of rocks giving pre-2,500 m.y. dates have also been found in a number of places in the Precambrian area of northern peninsular India from Calcutta extending westerly to Rajasthan.

All the Precambrian areas with rocks of this age show evidence of being involved in a major orogeny dating about 2,500 m.y. in which the volcanics and associated sediments were strongly folded, invaded by granites and sediments converted to granite gneisses.

2,500 to 1,800 m.y.

With the wearing down of the mountains formed by the massive orogeny mentioned above, new sediments, interspersed with lavas and ash beds, were deposited on the old basement. These formations were affected by another major orogeny dating about 1,900 to 1,800 m.y. and are exposed quite widely both in southern India and in the Precambrian area extending west of Calcutta.

The mineral deposits known to have been formed in this period include the major **iron**-producing belts of Singhbum and other parts of north-central India and their associated beds of **manganese** ore. The main Indian production of these metals comes from deposits of this age. **Copper** is also mined from the same general belts as is **uranium** from a small deposit at Jadaguda.

ORIGINAL ASIA

Pre-2,500 m.y.

The main area of early Precambrian in this part of Asia is in the eastern part of the People's Republic of China. Even here only a limited part of the Precambrian is exposed; a much larger "platform" covered by younger beds extends over most of China. Precambrian is also exposed in the cores of folds in the Himalayan and other systems of folding including the one that extends southerly from Burma into the Malaysian Peninsula.

In eastern China, rocks have been dated at 2,560 m.y. and formations of this general age probably exist in many of the other Precambrian exposures but successive later folding and metamorphism have made dating very difficult.

Mineralisation of this age probably occurs but is not known to form any major economic deposits.

2,500 to 1,800 m.y.

After the folding, faulting and intrusion dated about 2,600 m.y., erosion took place and sediments and volcanics were laid down on the erosion surface. In eastern China, these sediments and volcanics were affected by a period of folding and intrusion of granites dating about 1,800 m.y. The metamorphism was relatively weaker than that of the same age found in peninsular India.

Some low-grade **iron** ore beds in the People's Republic of China are probably of this age.

1,800 to 570 m.y.

Sediments, including quartzites, limestones and slates, lie unconformably on a basement of the rocks affected by the 1,800 m.y. orogeny both in Rajasthan and the areas east of it and in the Precambrian of southeast India. There was some further folding and metamorphism about 1,360 m.y. Following this, there was a very thick series of sediments deposited over a long period, from at least 1,200 m.y. to 550 m.y., i.e. into the early part of the Palaeozoic era. The lower beds of this series, known as the Vindhyan Series, are intruded by granites that date 735 m.y.

Although mineralisation is known in rocks of this period in various parts of India, none appears to be of major economic importance.

1,800 to 570 m.y.

In China and Mongolia there was deposition of marine sediments and volcanics and these were intruded by granites dating 1,320 m.y.

In the later stages of this time interval, there was formed a thick series of mainly marine, (but with some non-marine) sediments known as the Sinian System. They span the interval between Precambrian and Palaeozoic, being deposited from about 700 m.y. to 500 m.y., and in some parts of China these have remained undisturbed by later orogenies. They contain evidence of an ancient ice age at one stage.

From an economic standpoint, the system is interesting in that it includes **phosphate** beds grading up to 30% P_2O_5. This is the oldest known example anywhere of sedimentary phosphate of economic significance.

Sediments, highly altered but believed to be of late Precambrian age, are found in Laos and Vietnam.

570 to 225 m.y.

During this entire time interval from early Cambrian to late Triassic, India remained separated by 5,000 km from the main Asian continent and adjoined the east of southern Africa.

From the beginning of Cambrian time (about 570 m.y.) seas invaded the northern and eastern parts of the Precambrian craton depositing limestones and shales on the margins and probably on an extensive trench system that lay off the east side of the African continent. These are the sediments (shown on the map coloured dark blue) bordering the Precambrian areas of India and forming a major part of the Himalayan Range in a series of folds and thrust slices.

This marine sedimentation continued until about 350 m.y. when the African continent, of which "Original India" was a part, was elevated above sea level and a great series of dominantly terrestrial sediments were deposited in continental basins throughout a period lasting 150 million years. These deposits, called the "Gondwana Group", have a total thickness in India of over 7,000 metres. They include, near the base, rocks characteristic of an ice age which extended not only over India and Africa but also parts of South America, Australia and Antarctica, all of which formed part of Gondwanaland, the southern member of the Pangaea supercontinent. The glacial deposits were followed

570 to 225 m.y.

During the period from 570 to 340 m.y., seas covered the greater part of southern Asia and marine sediments are found in Iran, Afghanistan and Pakistan, and over most of China and Indochina. These sediments, in many places, are interbedded with submarine volcanic formations. By the beginning of the Carboniferous (345 m.y.) China was mainly above sea level and remained so for most of its subsequent geological history. Later sediments are dominantly terrestrial and include the oldest of the many **coal**-fields across the People's Republic. Iran and Afghanistan in the west and Burma in the southeast were still mainly below sea-level.

Volcanic activity was widespread across the continent, both submarine and on land, especially in Mongolia which is an area of remarkably frequent volcanic episodes throughout its geological history. In Pakistan a long period of volcanic activity started about 340 m.y., when the area was submerged, and continued in pulses, as the area rose above sea level, until about 200 m.y.

Intrusions of granites, and related rock-types, accompanied folding in several pulses during this time interval. In China the first major Palaeozoic granite episode seems to have occurred about 450 m.y. A second, and more extensive one, took place about 240 m.y. and its effect is seen on the map by a great series of mainly fresh-water sediments and include the main productive **coal** seams in India. Interlayered with the sediments are minor volcanics.

The greater part of peninsular India appears to have been elevated high enough to have missed the deposition of the Gondwana Group but it is present on strips along the east coast, in a northwest-trending belt centred about Latitude 18°N and Longitude 80°E and in a number of areas now north of Latitude 20°N.

as the broad curved belt that sweeps from east to west across northern China and Mongolia.

The area from Burma through Thailand, Malaysia and western Indonesia was involved in widespread granitic intrusion and it is with the latter that the world's most important **tin** province is associated. There appear to have been several surges of granitic intrusion as those of Malaysia give two distinct groups of dates, 300—280 and 230—175 m.y., while those in Thailand (also tin-bearing) date as young as 48 m.y. Production in 1979 from Thailand, Malaysia, western Indonesia and the southwestern corner of China, which forms part of the same "tin province" constituted 56% of the world total. This production came partly from the granite-associated vein mineralisation and partly from alluvial tin concentrates derived from the latter and transported into geologically-recent river and submarine sands and gravels.

While tin is the most important metal produced from rocks of this time interval in "Original Asia" other base metals, especially **lead** and **zinc** are locally important. An outstanding ore deposit with a long history of high grade lead-zinc-silver ore is the Bawdwin Mine in Burma which is in volcano-sedimentary rocks of about 500 m.y. age.

During the whole of this period from 570 to 225 m.y., petroleum products were probably formed widely in marine sediments but subsequent folding and faulting have largely dissipated the concentrations. One area where hydrocarbon production is from Palaeozoic beds is that of Yueng-Lengbu in north-central China.

220 to 65 m.y.

In the first part of this time interval India was still attached to Africa and the terrestrial Gondwana sediments continued to be formed on coastal and northern India until about 140 m.y. The central part of present peninsular India remained an elevated Precambrian platform uncovered by later sediments. This condition persisted after the rifting apart of India and Africa which occurred about 140 m.y. ago.

On the other hand, the larger submarine continental shelf which lay to the north of the Precambrian platform accumulated marine sediments (mainly limestones and shales) from early Triassic to late Cretaceous time, i.e. from before the parting and during the northward march of India until about 80 m.y. These marine sediments are found now from the Salt Range in the northwest, through the Himalayan region to northeast India.

220 to 65 m.y.

During this time interval there was, in the southern part of "Original Asia", an alternation of marine and non-marine sedimentation interlayered in places with volcanic material. In a broad sense, marine sediments predominate in the south and west while the proportion of continental sediments increases to the north and east.

The relative proportion of volcanic material increases in the same direction and is particularly abundant in Mongolia and northeast China. Some of the **iron** ores worked in the People's Republic of China are in volcanic related sediments of between 200 and 80 m.y. age.

In central Iran and Afghanistan marine sediments were deposited to a thickness of 2,500 metres giving way to continental sediments in the north.

As India approached Asia in late Cretaceous time, some folding and uplift of the northern continental shelf occurred and non-marine sediments, including some inferior **coal** beds, were deposited on the basement of marine beds.

About 75 m.y., the first outpouring commenced of the great basalt flows in western India known as the Deccan Traps. This volcanism continued into early Tertiary and the flows cover today, even after much subsequent erosion, over 500,000 km^2.

The continental sediments of this period in most of the area of this map-sheet include **coal** seams. These are mined especially in Iran and Pakistan and also to the east in the Malaysian Peninsula. But it is in the People's Republic of China where coal seams of this general age are most abundant. They form a large proportion of the huge reserves in that country, to a large extent still undeveloped.

A period of extensive granite intrusion took place in late Jurassic-early Cretaceous times, i.e. about 150 to 120 m.y ago. The largest area where these granites are exposed is in east China, in a belt along and parallel to the coast. Associated with some of these granites are the deposits of **tungsten**, of which the People's Republic produces 60% of the world's supplies, together with some **tin** and smaller amounts of other base metals. An **antimony**-producing field, of which metal China is also a major supplier, lies west of the tungsten-tin area.

Other parts of southern Asia having granitic intrusions of the same approximate age include Iran, Vietnam and Korea. At the same time, in Mongolia, there was a complex series of mafic as well as granitic intrusions that extended throughout most of the Mesozoic era.

ALL OF SOUTHERN ASIA

65 m.y. to Present

At the close of the previous time-interval "Original India" was approaching "Original Asia" with an intervening sea in which sediments, and some volcanics, were being deposited. The outpouring of basaltic lavas, the Deccan Traps, over western India continued until about 55 m.y. The actual collision appears to have occurred about 45 m.y. but the northward thrust of India continued, although at a slower pace, and it still continues. This movement is largely responsible for the widespread tectonic activity at present (and during Tertiary times) in the Himalayas and China. A brief study of the map brings out the immense folding, including the Himalayan Range, the great vertical movements (indicated by marine sediments elevated in a relatively short period to 5,000 metres above sea level) and the extensive faults with hundreds of kilometres of horizontal movement. The Altyn Tagh Fault and its extension, the Kansu Fault, form possibly the greatest active continental strike-slip plane in the world.

The tectonic activity up to the present day is also indicated by the relative frequency of major earthquakes. The most catastrophic earthquake in history was on January 23rd, 1556 near Hsian in Shensi Province where 830,000 people were reported killed. The most recent major quake, on August 6th, 1976 at T'angshan, evidently on a northeasterly-trending fault southeast of Peking, killed a reported 600,000 people. Since recorded history there have been over 90 earthquakes in China of magnitudes over 7 on the Richter scale, 80% of which were localised on recognised major faults. Of the 22 greatest earthquakes of the world between 1897 and 1955, seven ocurred in central and eastern China. The year 1976 appears to have been one of greater than usual earthquake activity in the world generally but especially in China where there were six quakes of over 7 on the Richter scale.

In the later stages of tectonic movement the Arabian Plate, which had split off from Africa along the Red Sea rift about

30 m.y. pushed against the Eurasian Plate and plunged under it in a subduction zone that dipped northeast under Iran-Afghanistan. This is defined in western Iran by the Zagros Thrust and can be followed to the southeast just off the coasts of Iran and Pakistan. At Longitude 66°E, this plate boundary (one of the youngest known) runs into the collision boundary of India with Asia which extends in an arc north-to-east parallel to the Indus valley.

The Arabian-Eurasian Plate boundary marks the northeastern limit of the area that contains over half the known world petroleum reserves (see notes on the Africa sheets). The greater proportion of the oil and gas was generated and emplaced before the collision of Arabia with Asia but the location of this remarkable accumulation in a sinking and narrowing arm of the sea between two converging continents is clearly of significance.

Returning to the time of the collision of India with Asia, seas continued to exist to the north of the advancing ram and on each side. Thus marine sediments continued to be deposited in Tibet and Pakistan to the north of the collision boundary, in Afghanistan and Iran to the west and in Burma to the east. However, with relatively rapid thickening of the crust, as a consequence of the collision and continuing northward movement of India, a general uplift was created in this part of Asia. Erosion became very vigorous and the marine basins rapidly filled and this, with the continued elevation, resulted in continental sedimentation replacing marine. This change is seen in Iran, Afghanistan and India in the form of red sandstones and salt beds. In west China and Tibet, continental sedimentation filled intermontane basins and fault-block depressions and the resulting monotonous sedimentary platforms are particularly noticeable on the map in the present Tarim and Zongar desert basins.

Outpourings of basalt and andesite lavas were common in these tectonically unstable times especially in Iran and Pakistan, in Mongolia and Manchuria and in the northern parts of Burma.

In relation to economic deposits, **hydrocarbons** were generated in the early stages of this time interval when marine sediments were being deposited. The great concentration in the vicinity of the Arabian/Persian Gulf is dealt with in the notes to the African sheets. In India, oil and gas are produced from beds of 60-40 m.y. on the west coast near Bombay, and in beds of 50-25 m.y. near the Burma border. In the People's Republic of China, the main oil and gas fields are in sediments of pre-65 m.y. but some, particularly in Szechuan Province, are in younger sediments.

In the Taching oil field, which was the largest producer in 1978-79, the source beds, unlike those of most of the oil fields of the world, are non-marine sediments.

With the change to terrestrial sedimentation, **coal** was formed in many parts of southern Asia including eastern India. A major part of the vast and widespread reserves of coal in China was formed earlier in Palaeozoic and Mesozoic times but some coal, and still more lignite beds, are dated between 40 and 15 m.y.

In metals there was a new episode, about 48 m.y., of **tin** mineralisation associated with granites, in the vicinity of the Malaysian Peninsula where the metal had been earlier deposited about 300 and 230 m.y. In recent geological history, tin ore derived from the above primary sources was removed by stream erosion and concentrated in river and offshore gravels.

Porphyry copper deposits, related to the subduction zone at the Arabia-Eurasia Plate boundary, occur in Iran, where the largest is the Sar Chesmeh deposit, dated about 12 to 15 m.y., and in Pakistan where three orebodies occur at Saindak, near the Iran border.

Chromite tends to be concentrated in ophiolites, the mafic rocks that are thrust up from the ocean floor during continental collisions. Examples of this are seen in Iran and the chain of ophiolites that mark the suture of the India-Asia collision is a potential source of chrome.

The great rivers of Asia, such as the Indus, Ganges, Brahmaputra, Irawaddy, Mekong, Yangtse Kiang and Huang Ho, have moved immense quantities of material from the highlands to inland flood plains and offshore deltas. The fact that some of these great rivers rise north of the Himalayas and cut through them in deep gorges indicates that they pre-date the uplift of these mountain ranges.

Sediments of recent age in Tibet and northern China include the loess deposits which are accumulations of wind-blown dust from the dried-out glacial debris at the terminal margins of glaciers that covered the higher mountain ranges in the Pleistocene Ice Age.

JAPANESE ISLANDS

The Japanese islands are part of one of the volcanic island arcs that fringe much of the Pacific Ocean. Thus the geological history, so far as observed effects are concerned, is relatively young and for this reason is summarised separately from the main Asian continent. Like most marginal areas along the edges of continents, however, these dominantly volcanic islands include terrestrial volcanics and basement of older metamorphosed sedimentary rocks. Thus gueisses of uncertain age and granites dating back as far as 430 m.y. are found in the south.

The basement of much of the later volcanic deposits that form the bulk of the islands is Permian (270 to 225 m.y.) and the main volcanic island arc processes appear to have commenced about 180 m.y. and continued in successive phases up to the present. Since much of the volcanic extruded material was submarine there is considerable interbedding of this with marine sediments. During periods of uplift, fresh-water sediments were also deposited.

About 50 m.y., the volcanic and sedimentary layers that had accumulated were involved in a period of folding and granitic intrusion, producing the metamorphic rock known as the "green tuff". This was followed by a period of subsidence and intense volcanic activity from 23 m.y. continuing for the following 10 million years. Younger granitic intrusions are dated at 13 m.y. Volcanic activity, but on a less widespread and violent scale, has continued over the last 10 million years up to the present.

Japan has a small but important mining industry. The oldest significant metallic deposits are of the skarn type, at or near the contacts of granitic intrusions, and 70% of the internally-produced **iron** in Japan comes from deposits of this type as well as about 35% of the **zinc**. The more important metallic deposits, however, are those of **copper, lead** and **zinc** (with accompanying **silver**) associated with the late stages of a volcanic period 13 m.y. These include layered deposits of the "Kuroko" or "black ore" type, plus massive pyrite and disseminated chalcopyrite, and vein deposits of the same metals.

Uranium has been found and developed in a considerable number of small deposits in non-marine sediments of 90 to 60 m.y.

Coal is found in non-marine sediments of about 60 to 50 m.y. and some 18 million tonnes a year are produced although this represents only a quarter of consumption in Japan.

Oil and **gas** are exploited from sediments younger than 35 m.y. but in limited quantities and the greater part of consumption depends on imports.

Japan had the fourth largest commercial use of geothermal power in 1977; a total of about 170 megawatts in five separate areas.

THE INDONESIA-AUSTRALASIA SHEET

This sheet covers four contrasting land groups:

(1) Australia, an ancient land mass that has a relatively large proportion of its area showing rocks undisturbed in the last 500 million years.

(2) New Zealand, originally a part of eastern Australia, which broke away between 80-60 million years ago.

(3) The belt of large and small islands extending from western Indonesia to New Caledonia with a generally young but very complex history.

(4) The Philippine Islands which are treated separately because their geological history differs from that of the Indonesian Archipelago.

In viewing the history of the Australasian continent, one must understand the concept, now widely accepted by earth scientists, that New Zealand was attached to eastern Australia until 80 to 60 m.y. and Australia did not part from Antarctica until 53 m.y. On the other side of Australia, the present west and north coastal areas (or the continental shelf beyond them) were probably attached to India-Tibet until 125 m.y. but there is less agreement, among those who have studied the matter, on the "fit" prior to dismemberment in this area than in almost any other part of the Earth.

Under present conditions, Australia and New Zealand both lie on the same "Indo-Australian Plate" while the Indonesia-New Caledonia Archipelago lies approximately along the north boundary of this plate which corresponds to a south-dipping subduction zone of geologically-recent age.

AUSTRALIA

Pre-2,600 m.y.

Australia has one of the largest areas of Archaean (pre-2,600 m.y.) rocks of the world and a large proportion of this is concentrated in the Yilgarn and Pilbara ranges of Western Australia. As in the older centres or cratons of Precambrian in other parts of the world, these consist mainly of two classes of rocks—relatively narrow belts of closely-folded volcanic formations with subsidiary related sediments, separated by wider areas of gneisses and granites. The volcanic ("greenstone") belts may be the remnants of much more widespread volcanic activity but it seems probable that this was concentrated to some degree along linear structures inherited from the initial continental crust. The intervening gneisses probably represent altered sandy or feldspathic sediments in part but may include parts of the primordial crust. In common with such early Precambrian areas in other parts of the world, those in Australia underwent a major orogeny—folding, faulting, intense alteration and intrusion by dominantly granitic magma—which occurred about 3,000 m.y. in the case of Pilbara and 2,600 m.y. in the Yilgarn.

Although the atmosphere at this time may have contained little oxygen, evidence has been found for forms of primitive life of the blue-green algae type, in rocks which date about 3,000 m.y.

Metallic deposits of economic importance were formed at the time of, and related to, the above volcanic activity. These include:-

(1) The **gold**-bearing veins in "greenstone belts" that played a major part in the opening up and settlement of Western Australia. The most productive areas, in the past and today, are those of Kalgoorlie and Norseman. A recent discovery is the Telfer Mine.

(2) The **nickel** deposits in ultra-mafic lava flows and sills in the same greenstone belts, especially in the Kambalda, Windarra and Agnew areas.

(3) **Copper-zinc** sulphide deposits of this age related to acid or intermediate volcanic activity are not as numerous as in the Archaean areas of Canada and southern Africa but examples are Golden Grove and Teutonic Bore.

2,600 to 1,600 m.y.

Following the 2,600 m.y. orogeny, and possibly coinciding with an increase in the oxygen content of the atmosphere, a period of extensive erosion occurred and water-lain sediments, with some volcanic flows and detritus, were deposited over and around these ancient, altered and worn-down cratons. The weathering of rocks gave sand and clay fractions that resulted in sediments much like those formed today, e.g. sandstone, shales and some limestones, but including extensive **iron** formation that is characteristic of this period worldwide. This phase of relatively quiet deposition ended in an orogeny that commenced about 1,800 m.y. affecting a large part of the present Australian mainland. Extensive exposures of the results of this orogeny are indicated on the map in the southern part of South Australia and western New South Wales, in the northern part of Western Australia, in the Northern Territory and northwestern Queensland. The strongly folded belt that forms an arc on the south side of the Kimberleys Basin appears to have been formed in this period.

Mineral deposits formed during the period prior to the 1,800 m.y. orogeny include the original deposition of the **iron** formations of the Hamersley Range (2,200 m.y.) but the concentration of iron to a grade suitable for direct-shipping ore probably occurred about 1,500 m.y. Also in rocks of this approximate age, and deposited during or following volcanic episodes about 1,680 m.y., are the **copper-lead-zinc** deposits of Mount Isa, the **zinc-lead** of McArthur River basin, the **copper-gold** of Tennant Creek and the unique composite orebody of **lead-zinc-silver** at Broken Hill, NSW. These were involved in a later orogeny between 1,700 and 1,400 m.y. before the present. The Mary Kathleen **uranium** deposit is in what were originally calcareous sediments, now granulites and gneisses, of the same general age. In the Northern Territory, **uranium** was deposited in certain sedimentary formations initially in sub-economic concentrations. The concentration to economic grade seems to have occurred during the following period.

1,600 to 900 m.y.

A third period of erosion and new sedimentation followed as indicated by exposures in the northern part of the Northern Territory and Queensland and in the northern part of Western Australia. These included sandy limes and shaly sediments interspersed, in the Bangemall Basin, by outpouring of lavas (1,100 m.y.). Folding and granitic intrusions occurred intermittently in different parts of the continent and especially east and south of the Yilgarn about 1,350 m.y. and later in local areas to the east.

As mentioned, the uraniferous formations laid down in the previous period were the source of **uranium** concentrations that occur in the eroded basement at the base of the sedimentary formations laid down about 1,700-1,600 m.y. These important concentrations include Jabiluka (possibly the world's largest known uranium concentration), the Ranger deposits, Nabarlek, Koongarra and the earlier-found Rum Jungle. Uranium concentrations of similar type were

formed in approximately the same period in the Precambrian of western Canada.

A discovery of importance was made in 1977 at Roxby Downs, South Australia where copper with significant associated uranium was found by drilling at the base of sediments (1,000 m.y.) lying on older granitised basement.

900 to 570 m.y.

This was a period lacking intense folding and encompassing the consolidation of the Australian Precambrian cratons as a whole. Cover sediments were formed on the older Central Platform areas and heavy deposition of sediments occurred in a north trending trough that was later involved in the Adelaide folding.

570 to 65 m.y.

Near the end of the Precambrian, and extending into the period of 500 m.y., a relatively minor orogeny with a northerly to northeasterly trend occurred in the south in a structure known as the "Adelaide Syncline". This resulted in folding of late Precambrian rocks and the early Palaeozoic sediments up to those dating about 450 m.y. Apart from this, the end of the Precambrian era was followed by an invasion of seas, with some periodic reversals, that deposited sediments dating 500 to 65 m.y. that now cover over half the continent, and probably at one time extended much further, but have since been partially eroded to uncover the Precambrian areas. In the later stages much of the sediment deposition was in fresh water bodies.

Much of this large expanse of sedimentary beds has remained relatively undisturbed in the central part of the continent. **Natural gas** and **oil** in limited quantities are found in sediments of about 490 m.y. in the Amadeus basin, southwest of Alice Springs, the oldest such occurrence in Australia. In the east, however, a great complex north-trending trench (the Tasman geosyncline) extended from Antarctica (then attached to Australia) through Tasmania and the east side of the Australian mainland and on to what is now New Guinea. It also involved the southwest side of New Zealand which was then also attached to the Australian mainland. Major trenches of this type off the edges of continents tend to subside for tens or hundreds of millions of years permitting the accumulation of huge thicknesses of sediments and some volcanics. They almost universally end in a great period of orogeny which, in this case, started about 470 m.y. on the western side of the geosyncline, reached a maximum about 200 m.y. and died out 100 m.y. It consisted, however, of a series of surges of folding, broken by periods of quiescence and sedimentation. In general, the more easterly fold belts, i.e. nearer the edge of the continent, are the younger. On the map, the lineated belts of blue (Palaeozoic) and green (Mesozoic) rocks, broken by granitic intrusions, show the results of this Tasman orogeny.

Most of the metallic deposits of the eastern side of the mainland and in Tasmania are in formations dating between 500 and 200 m.y. that were involved in the Tasman orogeny. Some of the mineral deposition occurred before the orogeny but some resulted from it. The Mount Lyell **copper,** Roseberry **lead-zinc** and Renison-Bell **tin** were formed during or following periods of volcanic activity prior to the folding as were many base metal occurrences in eastern NSW. (e.g. Cobar and Captains Flat that date 425 m.y.) and Queensland. Some, like the **copper-gold** at Mount Morgan, Queensland, are associated with intrusive plugs or volcanic pipes.

Climatic and topographic conditions suitable for bituminous coal formation occurred on several occasions during the interval from 250 to 140 m.y. seen in the coastal belt from NSW to Queensland where the Sydney and Bowen Basins are the most productive.

Hydrocarbon exploration on land in eastern Australia has had limited success but where production has been obtained or indicated it is mainly in sediments of Devonian (395-345 m.y.) age in Adavale, west Queensland and of Permian to Jurassic (280-136 m.y.) in other parts of the Eastern States. Major production however, is from younger beds off shore as mentioned below.

In the meantime in the west, beyond the broad central area that remained relatively undisturbed during the period from 570 m.y. to the present, a major structure developed in the Darling Fault and its related branches. This north-south structure resulted in a west-side-down movement which had the effect of preserving, on the western edge of the Australian mainland, sediments of 400 to 50 m.y. to a thickness of over 15,000 metres in some places. The vertical movement, which had probably started in 1,200 m.y. and continued in stages to 120 m.y., reached its maximum about 160 km north of Perth but dies out to the north and south and there was little if any horizontal movement.

No metallic deposits of economic importance have yet been found associated with this structure or the beds involved in it but they would be difficult to detect under the cover of younger beds.

The movements on the Darling Fault and branches from it may have been indirectly responsible for the accumulation of gas and oil on the northwest shelf of Western Australia. Up to the present, more gas than oil has been found and the reservoir beds are mainly of late Jurassic to early Cretaceous age.

The **manganese** deposits of Groote Eylandt in the Gulf of Carpentaria, which are mainly responsible for Australia holding the position of the world's fifth largest producer of the metal, are in marine sediments of late Cretaceous (about 75 m.y.) age but reworked in Tertiary times.

An event of interest in the study of climatic changes occurred in the period from 300 to 250 m.y. when at least two major ice ages affected the now-desert area straddling the Western Australia-South Australia boundary. Evidence of the same period of glaciation is seen in southern Africa, India, South America and Antarctica providing one of the initial observations that inspired the concept of the splitting up of formerly connected continents.

65 m.y. to Present

During most of the geological events referred to above, New Zealand had not parted from Australia nor the Australian mainland and Tasmania from Antarctica. While it is difficult to place accurate dates on the respective partitions the following indicates the approximate consensus of opinion of those who have studied the geological history of this part of the world:

130 m.y.—India parts from the west side of Australia and Antarctica.

80-69 m.y.—New Zealand parts from Australia.

53 m.y.—Australasia parts from Antarctica.

32 m.y.—Indian and Australian Plates, which were both moving north, joined and have been one plate since that time.

This combined Indo-Australian plate continued to move northwards and converge on the Eurasian Plate. The collision of Australia with the south-eastern part of the Eurasian Plate, between which deposition of sediments and lavas had been taking place, resulted in a complex south-dipping subduction zone in which the Australian continent overrode the younger folded volcanics and sediments on the Asian continental shelf. This event has occurred in the last 20 m.y. of geological history and is still in progress.

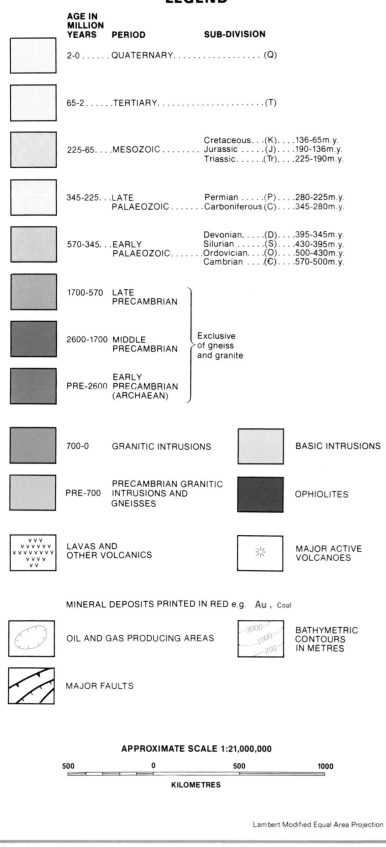

During the last 60 million years, the Australian continent has remained largely above sea level and subject to the effects on surface of air, water and temperature during successive changes in climate. The flat-topped hills of laterite, dissected sharply by later water erosion, are visible examples of such changes seen in now semi-desert areas. In the mountain areas of the southeast and in Tasmania, the scouring by glaciers and remains of moraines testify to the ice age that affected this part of the country and New Zealand and ended 12,000 years or so ago.

In the course of this prolonged exposure to surface effects a number of Australia's more important mineral deposits have been formed. Among the most significant of these are:

Bauxite (ore of aluminium) results from tropical or semi-tropical weathering processes acting on the surface of rocks having a relatively high alumina content. The most extensive deposits are in the northeast on the Cape York Peninsula. Other deposits, on a smaller scale, are in Western Australia south of Perth.

Heavy Mineral Sands primarily important for their rutile (titanium), ilmenite (titanium-iron), zircon (zirconium) and monazite (rare-earths) content were formed very recently, in fact are still being formed in places on the Queensland coast. In Western Australia, concentrations of the same type, but of less recent deposition, occur in raised beaches and river gravels inland from the coast.

Iron has been mechanically reconcentrated or redeposited in river valleys in Western Australia, for example on the Robe River, by the erosion of deposits of Precambrian age.

Uranium deposits of economic grade and size have been formed by solution, remobilisation and precipitation by seasonal surface waters under semi-desert conditions in Western Australia. The principal known example is at Yeelirrie where the uranium is in calcrete, a hard calcareous crust formed at the surface under arid conditions.

Some uranium concentrations in sediments in other parts of Australia, such as the Frome Basin of South Australia, have resulted from migration of uranium in solution in surface waters in recent geological times.

Gold, although initially in the oldest Precambrian rocks of western Australia and in Palaeozoic formations in southern and eastern Australia, has been reworked and concentrated as alluvial (placer) deposits in stream channels in the last period of Australia's geological history. It was the geologically recent concentration of alluvial gold, or its enrichment in near-surface deposits by the removal of more soluble barren accompanying minerals, that made possible the recovery of gold by individuals or small groups in the early gold-rush days.

In addition to the above deposits formed in the last 60 million years of Australia's geological history, many base metal deposits of Precambrian or Palaeozoic age, especially of **copper**, have been up-graded by the action of surface waters. These have leached the metals from near surface and precipitated them in more concentrated form lower down, especially at the lower limit of oxidation at the time. This secondary enrichment has, in some cases, made the difference between an uneconomic body and a grade that can be profitably mined on a small scale.

The formation of laterites, mentioned above, involves enrichment in iron at surface but normally not in a grade that can be exploited. When the underlying rock has an above-average nickel content, however, a **nickel laterite** of economic grade may result. An example is at Greenvale near the northeast coast. The classic nickel laterite occurrences are in New Caledonia as noted below.

Over 90% of Australia's production of **oil** (which supplies about 70% of the nation's consumption) and a large proportion of its reserves of oil and gas are in sediments of 65 to 40 m.y. in the Bass Straits area off the shore of the State of Victoria. The brown **coal** of Victoria, an important source of energy, is of Oligocene age (38 to 23 m.y.).

NEW ZEALAND

New Zealand, as mentioned previously, is believed to have been attached to Australia until 80 to 60 m.y. and thus shared the geological history of the eastern part of Australia from late Precambrian (say 1,000 to 570 m.y.) and the subsequent sedimentation in the great north-trending Tasman geosyncline. The sediments and volcanics that accumulated through Palaeozoic times became involved in the great orogeny of 200 to 100 m.y. that affected eastern Australia. In South Island of New Zealand, this orogeny, particularly active between 120 and 90 m.y., was the first of two events that produced the Southern Alps. After this first orogeny, the resulting mountains were almost completely eroded. Then, as recently as 10 m.y., a second major period of folding, faulting and uplift occurred forming the present spectacular Southern Alps. A major feature of this event is the Alpine fault, extending beyond the length of the two islands and involving "right-handed" horizontal movement (the northwest side moving northeastwards) to the extent of 500 km in addition to major but uneven vertical movements. In the meantime, volcanic activity, affecting especially the North Island but locally also the South Island, occurred in repeated phases from about 100 m.y. to the present time when its effects are seen in extensive hot spring activity.

Metallic mineral deposits have, to date, played only a small part in New Zealand's economy. **Gold**, in veins associated with volcanic episodes over the past 30 m.y., occurs in North Island and has been the source of alluvial deposits formed in geologically recent times. Alluvial gold in South Island, in Tertiary gravels in the Otago area and in recent river gravels in Westland (on the west coast) was derived from veins of a much older period, possibly pre-450 m.y. The vigorous erosion due to rapid elevation of the Southern Alps in late Tertiary time released the gold which, in most cases, has passed through several stages of alluvial concentration before reaching the present placer deposits. **Coal**, formed in the period 100 to 38 m.y., is mined near Huntly in North Island and near the northwest coast of South Island.

Iron-bearing sands of recent geological age off the coast of South Island are one of the most important mineral resources of New Zealand.

An event of recent geological age was the occurrence of an ice age in the last million years of geological history which covered the Southern Alps and adjoining areas as was the case in Tasmania and southern Australia. This corresponds in age approximately with the Pleistocene Ice Age of the Northern Hemisphere. Remnants of the ice cap are seen today as valley glaciers in the higher elevations of the mountains.

Perhaps the major mineral resource of New Zealand is **geothermal** energy, i.e. the tapping of steam and hot waters. It is commercially developed to date (1980) in only the Wairakei area on North Island but the potential for further development and use is substantial.

INDONESIA-NEW CALEDONIA BELT

The third unit on the map, the 8,500 km belt of large and small islands extending from western Sumatra to New Caledonia, has a generally younger but very complex history, the details of which are still relatively uncertain. Recent work suggests that the east and west portions of this great archipelago had very different histories. That part from Sumatra to west Sulawesi is thought to have been part of "Laurasia", the northern portion of the original grouping of continents that Wegener named "Pangaea", while the remainder of the chain from east Sulawesi and

Timor through New Guinea to New Caledonia was part of "Gondwana".

In the western portion of the archipelago, some accumulation of sediments appears to have been followed by one or more orogenies, including granite intrusions, from as far back as 280 m.y. but only little evidence of this remains uncovered by later sediments and volcanics. **Tin** deposits associated with granites in western Indonesia are described with those of countries to the north in the notes accompanying the Southern Asia sheet.

Timor, and the islands east of it, appear to have split off in front of a volcanic arc about 70 m.y. The main folding and accompanying volcanic activity in this area started at about 15 m.y. and resulted from the movement of the Indo-Australian Plate northwards against the Eurasian continent. This folding and volcanic activity is continuing to the present. This movement brought in line the originally separate parts of the present archipelago and caused the orogeny that follows the general east-southeast trend of the island chain now seen. The subduction zone that resulted caused the Australian continent to override the folded volcanics and sediments that had accumulated in the trench between it and Eurasia.

The orogeny involved granitic intrusions, some of which carry **copper** mineralisation in the form of porphyry copper deposits. Some of these are the youngest porphyry copper deposits known, being less than one million years old. Important examples are the producing mine in Bougainville, and the Ok Tedi deposit under development in Papua New Guinea. The Ertsberg deposit near the south coast of West Irian is not strictly a porphyry copper deposit but probably related to such.

New Caledonia has for many years been known for its reserves of **nickel**. These are mainly in the form of secondary concentrations resulting from the surface weathering, under tropical conditions, of ultrabasic rocks, largely material of ocean floor origin, in which nickel is distributed in subeconomic quantities. The resulting surface concentrations are laterites and, at their base, are higher-grade "garnierite" layers which have been the main source of New Caledonia nickel to date. Extensive nickel laterites are also being mined in Indonesia.

The volcanic activity along the whole Indonesia-New Caledonia belt has been intensive and is active up to the present. The eruption of Krakatoa, at the southeast tip of Sumatra, in 1883 competes with Tambora, some 1,300 km to the east for the position of being the largest and most destructive volcano in modern history.

Hydrocarbon production from young sediments is already an important economic feature in several parts of Indonesia and the potential of the off-shore areas of the whole geological belt for oil and gas production is high.

THE PHILIPPINE ISLANDS

The Philippine Islands are an example of an island arc that has developed on the ocean floor by several successive geologic events. The earliest cycle of such events started in the late Palaeozoic and terminated in a stable platform position by the end of the Permian (225 m.y.) when volcanic activity occurred interspersed with sedimentation. These volcanic and sedimentary strata were subsequently involved in a period of folding and moderate metamorphism followed by erosion and subsidence below sea level. Remnants of these older volcanic and sedimentary formations have been exposed by erosion on several of the major islands, particularly on the west side of the group.

The second major cycle started with widespread island arc vulcanism in the mid-Cretaceous (100 m.y.) and continued in several distinct phases to the present. Again the volcanics were interspersed with sedimentary deposits. On several occasions ocean floor material (ophiolites) was thrust up onto the volcanic-sedimentary accumulations and in some cases was later folded with them. As is often the case, the ophiolites carry **chromite** which is concentrated in places into bodies of commercial significance.

Other mineral deposits include:-

(1) **Copper,** in places associated with **zinc** and **lead,** has been deposited, probably mainly as submarine depositions, associated with the volcanic outpourings.

(2) Granitic intrusives associated with the volcanic activity carry bodies of "porphyry copper" in several locations.

(3) **Gold** and **silver** deposits, mainly in veins, are also associated with the younger volcanic centres.

The Antarctic continent covers 14,000,000 km^2 and, if placed over North America with the South Pole on Denver, Colorado, would extend from Miami to San Francisco and from just north of Mexico City to the shore of Hudson Bay. Some 98% of this area is covered by a continental ice sheet of 2,000 metres average thickness, in places as much as 4,500 metres. One-third of the underlying surface lies below sea level but this is partly due to the continent being depressed an average of 600 metres by the ice load. If the ice were removed, west Antarctica would be an archipelago of mountainous islands, separated by a marine basin from the larger eastern shield area.

Most of the ice surface lies between 2,000 and 4,000 metres above sea level and is penetrated in places by peaks of mountain ranges that reach heights of as much as 5,000 metres above sea level. The mean height of the whole continent is nearly three times that of all other continents.

During the Precambrian era, and for 400 million years after it, Antarctica was part of Gondwana and was attached to the southern edges of South America, Africa and Australia and to the east side of peninsular India. These connections maintained until about 180 m.y. when the South Atlantic started to open up as part of the general break-up of Gondwana. This partition, by about 100 m.y., included the southerly movement of Antarctica from Cape Horn, which may have been overlapped by the Antarctic Peninsula. At about the same time, Antarctica parted from Africa but remained attached to Australia until about 53 m.y.

Today, Antarctica shows three dominant geological units, based on the limited rock exposures on the edge of, or projecting through, the ice sheet. These are (i) A Precambrian Shield or Platform occupying the easterly two-thirds of the continent; (ii) The Trans-Antarctic Mountain Range, of Palaeozoic age, west of the Shield and extending from coast to coast in a northerly direction; (iii) A younger, northwest-trending, mountain range following the west coast and the Antarctic Peninsula.

The Precambrian Shield, where exposed, consists mainly of granites and gneisses, at least partly of Archaean age but probably affected by orogenies dating about 2,600 and 1,700 m.y. respectively. Younger Precambrian sediments and volcanics were deposited on the west side of the Shield and were involved in the same orogeny that affected the overlying early Palaeozoic rocks.

Following the close of the Precambrian era, Palaeozoic sediments and volcanics were deposited on the western edge of the Precambrian Shield and in a probable ocean trench west of it. These were involved, together with the late Precambrian basement on which they lay, in a major orogeny that formed the Trans-Antarctic Mountains. The axis of this range roughly follows Longitudes 20°W and 160°E and the dating of granites that accompanied the orogeny indicates that it took place between 500 and 440 m.y. The erosion of the Trans-Antarctic Mountain Range in the succeeding period resulted in sedimentation between the east flank of the range and the Precambrian Shield. Sediments here range in probable age from approximately 350 to 150 m.y.

About 180 m.y., i.e. about the time of the start of the Cordilleran orogeny in South America and possibly an extension of it, an orogeny started to affect Palaeozoic and some early Mesozoic sediments and volcanics in western Antarctica. Folding and intrusion probably continued to about 90 m.y as indicated by granitic exposures along the west coast and in the Antarctic Peninsula.

Some sedimentation has taken place since the last orogeny on the narrow continental shelves surrounding Antarctica and in embayments such as the Wedell and Ross Seas. The last igneous event was volcanic activity on the west coast in the late Tertiary time (prior to the formation of the ice cap) which has continued to the present as indicated by the dying activity of the cone of Mt. Erebus on the east side of the Ross Sea.

As yet there are no producing mineral deposits in Antarctica and, although quite a number of mineral deposits have been noted, none are likely to be economically viable under the present conditions of supply and demand with the possible exception of oil and gas. There is evidence of quite

ANTARCTICA

LEGEND

AGE IN MILLION YEARS	PERIOD	SUB-DIVISION	
2-0	QUATERNARY	(Q)	
65-2	TERTIARY	(T)	
225-65	MESOZOIC	Cretaceous (K) .136-65 m.y.	
		Jurassic (J) .190-136 m.y.	
		Triassic (Tr) .225-190 m.y.	
345-225	LATE PALAEOZOIC	Permian (P) .280-225 m.y.	
		Carboniferous (C) .345-280 m.y.	
570-345	EARLY PALAEOZOIC	Devonian (D) .395-345 m.y.	
		Silurian (S) .430-395 m.y.	
		Ordovician (O) .500-430 m.y.	
		Cambrian (€) .570-500 m.y.	
1700-570	LATE PRECAMBRIAN		
2600-1700	MIDDLE PRECAMBRIAN	Exclusive of gneiss and granite	
PRE-2600	EARLY PRECAMBRIAN (ARCHAEAN)		
700-0	GRANITIC INTRUSIONS		
PRE-700	PRECAMBRIAN GRANITIC INTRUSIONS AND GNEISSES		

MAJOR ACTIVE VOLCANOES

LAVAS AND OTHER VOLCANICS

BATHYMETRIC CONTOURS IN METRES

MINERAL DEPOSITS PRINTED IN RED e.g. Au , Coal

APPROXIMATE SCALE 1:30,000,000

500   0   500   1000

KILOMETRES

Polar Projection

extensive iron formations in the Precambrian Shield of eastern Antarctica but it is unlikely that these can be worked, even where not ice covered, in competition with the extensive and more cheaply-worked iron ores on other continents.

From indications to date, base metal and precious metal deposits in Antarctica are more prevalent in rocks of Palaeozoic (570 to 290 m.y.) age than they are in Precambrian rocks of the eastern part of the continent.

Of potential interest is the existence of the Dufek layered igneous complex, inland from the Ronne Ice Shelf. This complex, covering 34,000 km^2, has similarities to the rocks of the Bushveld complex and thus has potential interest for nickel, chromium and platinum group metals.

The Earth's poles have "wandered" widely relative to the continents during geological history but, for the past 50 million years or so, the Arctic polar region has been occupied mainly by ocean crust bordered by the northern edges of North America, Greenland and Eurasia. This part of the crust has had less geologically recent tectonic change than regions nearer the equator. The mid-Atlantic ridge, along which the Americas parted from Eurasia and Africa, passes through Iceland (a "hot spot" of activity on the ridge) and continues along the northeastern side of Greenland but swings easterly and passes 450 km east of the pole. It heads southerly along Longitude 130°E before dying out as a recognisable structure near the USSR northern coast. Continental parting on this structure decreases north of Iceland and the total divergence near the pole appears to have been quite small.

Bathymetric surveys, since 1976 in the Arctic polar regions, have shown much more variation in depth than had been realised. The deep basins, with roughly parallel long axes, have been named the Canada, Makarov, Fram and Nansen Basins and are depressions in the ocean crust as marked on the map. The Lomonosov Ridge, separating the Makarov and Fram Basins, is believed to be a torn-off fragment of continental crust and has steep sides that are probably fault-controlled.

The geological history of the continental areas bordering the Arctic polar region is summarised in the notes to the respective map sheets. It may be noted that the Caledonian folding appears to split into two branches. That on the east side of Greenland swings west along the northern edge of that sub-continent while the Caledonian folding of Scotland and Scandinavia swings easterly. The Ural Mountains' folding that runs north approximately on Longitude 60°E swings westerly in an "S" shaped bend through Novaya Zemlya and then continues north easterly. Thus all fold systems that have a northerly factor to their orientation tend to swing more to latitudinal orientation as they approach the pole.

Mineral deposits within the 66°32'N Latitude of the Arctic Circle are more explored and developed in Eurasia than in North America. In Alaska there is no active mining, other than placer gold, north of the Arctic Circle. In Canada, copper of late Precambrian age was known near the mouth of the Coppermine river before the arrival of Europeans but has never been developed on a modern commercial basis. The only producing mine (zinc and lead) is at Nanasivik (Latitude 73°N) on the northwestern tip of Baffin Island.

A potentially profitable zinc-lead deposit is known in Ordovician sediments on Little Cornwallis Island (Latitude 74°50' N) and other deposits of similar mineralisation are known on this and neighbouring islands in a "hinge-zone" that will probably be traced through the east-central part of Ellesmere Island. A very high grade iron deposit, and extensive lower grade iron formations, are known in Baffin Island at Latitude 71°N, Longitude 78°W but abundant sources available in more accessible parts of the world have delayed the development of these or the extensive iron formations on the Melville Peninsula to the west.

Coal has been found at many places in Cretaceous sediments in Ellesmere Island but the energy that would be required to mine and ship it could exceed the energy it would produce when burned, so it will probably be restricted to local use when such demand occurs.

Oil and gas, both on land areas and offshore, constitute by far the greatest mineral potential in the North American Arctic. The reservoir beds are mainly between 180 m.y. and 90 m.y. in age.

Greenland has the disadvantage of having 80% of its 2,186,000 km^2 area covered by the last remaining Arctic continental ice sheet of significant size from the Pleistocene Ice Age. It is probable that many Precambrian and Palaeozoic ore deposits lie under the ice, some of which may eventually be located by geophysical techniques. One lead-zinc mine in Precambrian limestone at Marmorilik on the west coast at Latitude 71°N is in production.

In the Scandinavian Arctic, metal mining has a long history and the more important mineral areas are shown on the Northern Europe sheet. Similarly in USSR, north of the Arctic Circle, mining development is considerably more advanced than in corresponding latitudes in North America.

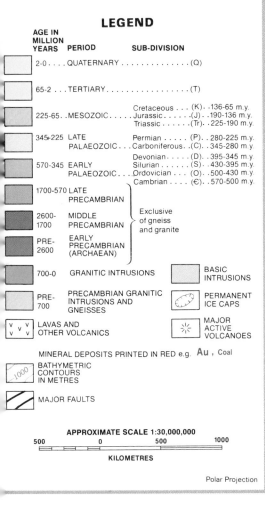

WORLD MINERAL PRODUCTION AND RESERVES

To give some idea of the global distribution of the principal minerals and metals of economic significance, the following tables of current production and reserves are presented. The notes also give, in very summarised form, the relative importance of formations of different geological type and age as hosts for particular minerals.

The most complete production figures available at the time of publication are those for 1978. Some official figures for 1979 have been obtained and estimates have been used for the remainder. The production figures will, of course, be rapidly out-dated but the *percentages* of global output from individual nations are also given and these change relatively gradually. To obtain more up-to-date figures, subsequent to those presented, the reader is referred to *Mineral Commodity Summaries* which is published annually (about February) by the US Bureau of Mines, and *Mineral Trade Notes* issued monthly or bi-monthly by the same office. Much of the data used here is derived from the above two publications, augmented by the cooperation of two departments of the Institute of Geological Sciences, London, and the corresponding departments of the Canadian Government in Ottawa. In a few cases, changes from these sources have been made on the basis of personal contacts or knowledge.

In the case of the figures on coal these were obtained from the publication *World Coal*, November 1979 edition. Figures on oil and gas are, to a large extent, from the *Oil and Gas Journal*, December 25, 1979.

When it comes to reserves figures, one is in comparatively uncharted waters for several reasons. In the first place, if the price of a metal or mineral remains unchanged for several years in a period of general worldwide inflation the "reserves", in the sense of material that can be produced at a profit, are bound to be rather drastically reduced. On the other hand, if the price of a metal or mineral makes a series of rapid increases (such as occurred in 1978 and 1979 in oil, gold, silver and platinum) material that was too low-grade or too difficult to recover to be regarded as reserves suddenly comes into that category. In practice, the various government officials usually wait

for a year or more before deciding that the increased price is likely to be permanent and that the reserves justify re-calculation on this account.

The other main reason for the unreliability of reserves figures presented by government officials is that few of them use the same parameters in establishing what are to be classed as reserves. This, naturally, affects the figures assembled by the U.S. Bureau of Mines which is probably the most up-to-date global gathering centre. Some countries do not release reserve figures at all. Some, including the USA, hold back on certain minerals considered to be of strategic significance.

Many books have been written on the intricacies and limitations of the records of world mineral reserves. One of the best, in the 1976-1979 period, is *World Mineral Supplies*, edited by G. S. J. Govett and M. H. Govett, published by Elsevier Scientific Publishing Co. in 1976. Most responsible writers in this and other recent publications agree that orders-of-magnitude on most mineral reserves are possible but that it is meaningless to regard them as specific, fixed amounts to relate against annual world consumption. This latter procedure was used, with very misleading results from the minerals standpoint, by the team commissioned by the Club of Rome in their book *The Limits to Growth*, published in 1972 by Universe Books and using reserves figures from the US Bureau of Mines of 1970.

The incompleteness and inadequacies of the available reserves data made the present authors doubtful of the value of including them at all in this publication but it was finally decided that figures giving at least an order-of-magnitude for comparative purposes would be generally useful. Where official figures are missing, the gaps have been filled by estimates from the best available sources.

The following list is a summary of the commodities for which production and reserve figures have been included in this section:

Non-Metallics	Metals		Energy Minerals
Asbestos	Antimony	Manganese	Coal
Diamonds	Bauxite	Molybdenum	Natural Gas
Phosphate	Chromite	Nickel	Oil
Evaporites	Cobalt	Platinum Group	Uranium
Potash	Copper	Silver	Geothermal Energy
	Gold	Tin	
	Iron	Titanium	
	Lead	Tungsten	
		Zinc	

NON-METALLICS

ASBESTOS

Crysotile asbestos, the predominant variety used in industry, is mined exclusively from ultra-mafic rocks, either intrusive or volcanic. In many of the important producing areas, the host rock is now believed to be ophiolite, i.e. ocean floor material interfolded with sediments during plate tectonic movements. In the case of Quebec, one of the most productive areas, the ophiolites were infolded about 350 m.y. ago. Southern African production, and some Canadian, is from early Precambrian rocks.

Country	Production 1978		Production 1979*	1979 Reserves*
	Tonnes	% of World	Tonnes	Thousand Tonnes
(1) USSR	2,500,000	48.75	2,600,000	50,000
(2) Canada	1,379,000	26.89	1,300,000	100,000
(3) South Africa	257,325	5.02	250,000	24,000
(4) Zimbabwe	225,000	4.39	230,000	20,000
(5) China (PRC)	220,000	4.29	200,000	50,000
(6) Italy	135,402	2.64	150,000	2,000
(7) USA	93,097	1.82	93,000	10,000
(8) Brazil	92,800	1.81	95,000	n.a.
(9) Australia	55,000	1.07	55,000	2,000
(10) Cyprus	37,857	0.74	38,000	2,000
(11) Swaziland	36,951	0.72	38,000	n.a.
Other	95,815	1.87	96,000	n.a.
TOTAL	5,128,247	100.00	5,145,000	300,000

* estimate

Note: Production and reserve figures expressed as asbestos fibre.

n.a. Not available.

Unless stated otherwise, all production and reserves figures are quoted in tonnes (metric tons) of 2204.62lb

DIAMONDS

Both gem and industrial diamonds are produced exclusively from pipes or dykes of kimberlite, an ultra-mafic rock of mantle origin, or from river or beach gravels in which the diamonds were derived by weathering of these kimberlite sources. Kimberlite intrusives are found in old cratonic areas and tend to come in surges but over a wide range of time. The majority of diamond-bearing kimberlites in southern Africa are of about 180 m.y. age, but some are over 1,000 m.y. In the USSR, the growing diamond industry in northeastern Siberia is from kimberlites dated about 290 to 250 m.y.

Production statistics of gem diamonds based on weight are not meaningful due to the great variation in value depending on the size of individual stones. The probable order of gem production by value in 1978 was:—Republic of South Africa, USSR, Namibia, Zaire, Botswana, Angola.

Country	1978 Production Gem + Industrial		Production 1979* Gem + Industrial	1979 Reserves[1] *
	Thousand Carats	% of World	Thousand Carats	Thousand Carats
(1) Zaire	17,470	36.62	21,500	500,000
(2) USSR	12,500	26.20	15,200	25,000
(3) South Africa	7,390	15.49	9,000	50,000
(4) Botswana	2,780	5.83	3,000	50,000
(5) Namibia	1,900	3.98	2,000	25,000
(6) Ghana	1,480	3.10	1,500	15,000
(7) Sierra Leone	810	1.70	1,100	6,000
(8) Venezuela	790	1.66	950	3,000
(9) Angola	650	1.36	900	20,000
(10) Liberia	520	1.09	700	2,000
(11) Brazil	480	1.01	650	5,000
(12) Tanzania	350	0.73	350	2,000
(13) Central African Rep.	290	0.61	300	500
(14) Ivory Coast	140	0.29	150	500
(15) Guinea	90	0.19	100	200
Other	70	0.14	100	500
TOTAL	47,710	100.00	57,500	704,700

* *estimate*

(1) Figures include both placer and kimberlite deposits.

PHOSPHATE

Although apatite and other phosphatic minerals are recovered from igneous rocks in various parts of the world over 80% of the phosphates supplying the fertiliser industry is from bedded rocks deposited on shelf areas on the edges of continents, usually with a high proportion of marine life. The origin of the phosphatic material is probably from the solution in seas or brackish water of decomposition products of dead organisms and the precipitation of phosphates by chemical or bacterial action. Such action takes place more readily in warm waters and most major phosphate beds were probably formed in fairly low latitudes.

Over 55% of present world production is from beds less than 100 m.y. but large reserves exist in western USA dated about 260 m.y. and important deposits occur in sediments dating between 700 and 500 m.y. especially in China and USSR.

Country	Production 1978		Production 1979*	1979 Reserves*
	Thousand Tonnes	% of World	Thousand Tonnes	Million Tonnes
(1) USA	50,037	40.36	51,342	8,500
(2) USSR	24,000	19.36	24,500	11,000
(3) Morocco	19,278	15.55	20,000	40,000
(4) China (PRC)	4,300	3.47	4,500	1,000
(5) Tunisia	3,712	2.99	3,850	280
(6) Togo	2,827	2.28	3,000	50
(7) South Africa	2,699	2.18	3,100	60
(8) Jordan	2,223	1.79	2,500	550
(9) Nauru Island	1,999	1.61	2,000	50
(10) Senegal	1,843	1.49	1,800	130
(11) Israel	1,759	1.42	1,900	500
(12) Vietnam	1,600	1.29	1,600	100
(13) Christmas Island (Indian Ocean)	1,386	1.12	1,400	50
(14) Algeria	1,136	0.92	1,250	500
(15) Brazil	1,094	0.88	1,300	400
Other	4,089	3.30	4,500	11,830
TOTAL	123,982	100.00	128,540	75,000

*estimate

EVAPORITES

Under the term "Evaporites" are included all the commonly used minerals that result from the concentration of salts in sea water to the point where individual minerals precipitate successively or in cycles. These include common salt (sodium chloride), gypsum and anhydrite (hydrous and anhydrous calcium sulphate respectively), barite (barium sulphate) and the various potassium minerals. The latter are reviewed separately since the major production comes from relatively few countries. The other salts, however, are very widely distributed, relatively low cost and, therefore, transported less from country to country. For these reasons no listing of production and reserve figures are given.

Evaporites have formed to some extent since about 1,500 m.y. although the sea at that time is believed to have had a much lower content of salts. The major geological periods of evaporite formation were during middle Precambrian (Helikian); middle Palaeozoic (Ordovician-Devonian); late Palaeozoic (Permian) and Tertiary (Miocene) times.

POTASH

Commercial potash, in the form of several different minerals, occurs in sediments formed as a result of the increased salinity and selective precipitation that occur when arms of the sea become periodically cut-off by vertical crustal movements. Consequently potash tends to be interbedded with common salt and gypsum.

Commercial production is mainly from sediments younger than 400 m.y., major examples being in Saskatchewan, Canada, where the containing beds are dated in the range of 370 m.y.; New Mexico, USA, 250 m.y.; France and Germany, 230 m.y. Large reserves of about 100 m.y. exist in Thailand and Laos, in eastern Brazil and in the Congo. Some supplies, however, are from modern brines and salt lakes such as the production by Israel from the Dead Sea and that from the Great Salt Lake area in Utah, USA.

Country	Production 1978		Production 1979*	1979 Reserves*
	Thousand Tonnes K_2O	% of World	Thousand Tonnes K_2O	Million Tonnes K_2O
(1) USSR	8,200	31.45	8,400	3,000
(2) Canada	6,124	23.49	6,720	2,700
(3) Germany (DDR)	3,320	12.73	3,400	500
(4) Germany (FDR)	2,470	9.47	2,500	500
(5) USA	2,253	8.64	2,120	300
(6) France	1,795	6.88	1,900	50
(7) Israel	691	2.65	720	300
(8) Spain	615	2.36	650	60
(9) China (PRC)	300*	1.15	300	100
(10) United Kingdom	150	0.58	300	50
(11) Italy	139	0.53	140	10
(12) Chile	17	0.07	20	10
TOTAL	26,074	100.00	27,170	7,580

* estimate

Note: Available plant nutrient (fertiliser) in the various potash salts produced is expressed in terms of K_2O equivalent.

METALS

ANTIMONY

Deposits of this metal that are of economic grade and quality are usually of relatively young age. The world's largest producer is China (PRC) where it appears to come from a large number of deposits associated with acid intrusives of 120 to 150 m.y.

Country	Production 1978		Production 1979*	1979 Reserves*
	Tonnes	% of World	Tonnes	Tonnes
(1) Bolivia	12,672	18.94	14,500	360,000
(2) China (PRC)	12,000	17.94	12,000	2,200,000
(3) South Africa	10,478	15.66	11,000	320,000
(4) USSR	7,900	11.81	7,900	270,000
(5) Canada	3,000	4.48	3,000	60,000
(6) Thailand	2,873	4.29	3,000	90,000
(7) Yugoslavia	2,760	4.13	2,300	90,000
(8) Mexico	2,457	3.67	2,700	220,000
(9) Turkey	2,440	3.65	3,000	110,000
(10) Morocco	2,110	3.15	2,000	100,000
(11) Australia	2,100	3.14	2,000	100,000
Other	6,116	9.14	8,500	430,000
TOTAL	66,906	100.00	71,900	4,350,000

* estimate

BAUXITE

Although aluminium is one of the commonest elements in the continental crust, the oxide, alumina, is recovered economically only from bauxite, a rock composed of one or more aluminium hydroxide minerals with a variable quantity of impurities. In times when bauxite is unavailable, e.g. in World War II when sea freight was interrupted, or in places where it is scarce, alumina has been recovered from clays and aluminous igneous rocks but not yet at competitive cost. Part of the USSR alumina supplies are from non-bauxite sources.

Bauxite is formed by the weathering of sediments or igneous rocks having relatively high alumina contents. The process is effective only under tropical or subtropical climates and, consequently, bauxite formed on the present surface is restricted to low latitudes as highlighted in the list below. Bauxite has, however, been formed on old surfaces in past geological periods and such "fossil bauxites" may occur at any latitude but are likely to be covered by later sediments and thus more expensive to mine. A "fossil bauxite" is found in southern Europe where it was formed on a surface dated about 130 m.y. and is mined in several countries, including Greece and Yugoslavia.

Country	Production 1978		Production 1979*	1979 Reserves*
	Thousand Tonnes	% of World	Thousand Tonnes	Million Tonnes
(1) Australia	24,300	29.14	26,000	4,600
(2) Jamaica	11,736	14.07	11,800	2,000
(3) Guinea	11,000	13.19	12,500	6,500
(4) USSR	6,180[1]	7.41	4,600	300
(5) Surinam	5,200	6.23	5,000	500
(6) Guyana	3,475	4.17	2,500	700
(7) Hungary	2,898	3.47	3,000	300
(8) Greece	2,630	3.15	2,600	700
(9) Yugoslavia	2,566	3.08	2,800	400
(10) France	1,978	2.37	2,000	n.a.
(11) USA	1,670[2]	2.00	1,700	40
(12) India	1,600	1.92	1,800	n.a.
(13) Brazil	1,400	1.68	2,500	2,500
Other	6,770	8.12	6,200	n.a.
TOTAL	83,403	100.00	85,000	22,700

* estimate   n.a. Not available.

(1) Includes 1,580,000 tonnes of bauxite equivalent of other aluminous rock.

(2) Dry bauxite equivalent of crude ore

CHROMITE

Economic deposits of the mineral chromite are almost universally associated with mafic or ultra-mafic rocks. These may be in the form of intrusions of deep, probably mantle, origin such as the "Great Dyke" of Zimbabwe and the mafic part of the Bushveld Complex of South Africa. Alternatively, the mafic rock may be in the form of ophiolites which are of ocean floor origin and have been transported tectonically onto continental rocks especially at plate margins, for example, Turkey, Albania and the Philippines.

Country	Production 1978		Production 1979*	1979 Reserves*
	Thousand Tonnes	% of World	Thousand Tonnes	Thousand Tonnes
(1) South Africa	3,145	32.85	3,180	1,400,000
(2) USSR	2,300	24.01	2,500	15,000
(3) Albania	930	9.71	900	2,000
(4) Turkey	680	7.10	680	5,000
(5) Zimbabwe	600	6.25	600	1,000,000
(6) Philippines	532	5.55	540	3,000
(7) Finland	507	5.29	530	25,000
(8) India	266	2.78	280	5,000
(9) Brazil	190	1.98	200	2,000
(10) Iran	163	1.70	130	2,000
(11) Malagasy Republic	138	1.44	140	2,000
Other	130	1.36	120	90,000
TOTAL	9,580	100.00	9,800	2,551,000

*estimate

COBALT

This metal is recovered almost exclusively as a by-product from other metallic operations especially those of copper and nickel. Sedimentary copper deposits in Zaire are particularly high in by-product cobalt. A few specific cobalt mines exist including one in Morocco.

When the mining of "manganese nodules" from the ocean bottom becomes feasible, there will be a substantial production of cobalt from this source, since this is the fifth most abundant contained metal, after manganese, iron, nickel and copper.

Country	Production 1978		Production 1979*	1979 Reserves*
	Tonnes	% of World	Tonnes	Tonnes
(1) Zaire	11,000	35.66	13,150	1,200,000
(2) New Caledonia	4,170	13.53	4,170	91,000
(3) Australia	3,450	11.20	3,450	45,000
(4) Zambia	2,270	7.36	2,700	360,000
(5) USSR	1,950	6.32	2,000	91,000
(6) Canada	1,800	5.83	900	27,000
(7) Morocco	1,800	5.83	1,800	14,000
(8) Cuba	1,600	5.19	1,600	n.a.
(9) Finland	1,360	4.41	1,400	18,000
(10) Philippines	1,100	3.56	1,100	180,000
Other	350	1.13	230	n.a.
TOTAL	30,850	100.00	32,500	3,000,000

*estimate
n.a. Not available.

COPPER

Under present conditions, over 50% of copper production throughout the world comes from "porphyry" type deposits varying in age from 200 m.y. to less than 1 m.y. These are relatively low grade, usually mined by openpit methods and are found especially along the Cordilleras of North and South America, but also in the western Pacific and in Iran and adjoining Pakistan. This trend towards large, low-grade deposits as the major source of the metal is comparatively recent. The next most important producing class is that of bedded copper of late Precambrian age (about 1,000 m.y.) as found in Zaire and Zambia. The third important source is massive sulphide deposits, associated with volcanic activity, especially of pre-2,500 m.y. age, but also in younger formations.

Country	Production 1978		Production 1979*	1979 Reserves*
	Thousand Tonnes	% of World	Thousand Tonnes	Thousand Tonnes
(1) USA	1,358	18.13	1,430	101,000
(2) Chile	1,036	13.83	1,040	107,000
(3) USSR	865	11.55	840	40,000
(4) Canada	658	8.79	620	35,000
(5) Zambia	643	8.59	590	37,000
(6) Zaire	423	5.65	380	26,000
(7) Peru	366	4.89	370	35,000
(8) Poland	290	3.87	300	14,000
(9) Philippines	239	3.19	300	20,000
(10) Australia	220	2.94	230	9,000
(11) South Africa	209	2.79	190	5,000
(12) China (PRC)	200	2.67	200	5,000
(13) Papua New Guinea	199	2.66	180	16,000
(14) Yugoslavia	121	1.62	120	n.a.
Other	662	8.84	590	n.a.
TOTAL	7,489	100.00	7,380	550,000

*estimate
n.a. Not available.

GOLD

While much of the former, and some current, production is from alluvial (placer) deposits of recent origin, the primary source of gold production and reserves is predominantly in Precambrian rocks of pre-2,500 m.y. age. South African production is from conglomerates dating from 2,600 to 2,200 m.y., but the source of the gold is earlier Precambrian rocks. Much of the production of USSR is also of Precambrian original source, but 55% is currently obtained from alluvial deposits.

A second period of gold mineralisation is much younger, mainly in vein-type deposits of Tertiary age. Examples are the veins in Cordilleran folded rocks which supplied the gold that attracted the Conquistadors to Central and South America and contributed much of the incentive to the opening up of western USA.

Approximately 20% of current gold production is in the form of by-product from base metal operations, especially copper.

Country	Production 1978		Production 1979*	1979 Reserves*
	Tonnes†	% of World	Tonnes	Tonnes
(1) South Africa	706.4	51.24	703	16,500
(2) USSR	410.0	29.73	430	15,500
(3) Canada	52.9	3.84	49	1,550
(4) USA	30.2	2.19	28	1,400
(5) Papua New Guinea	23.4	1.70	20	390
(6) Australia	20.2	1.46	20	250
(7) Philippines	17.2	1.25	30	60
(8) Zimbabwe	17.0	1.23	15	140
(9) Ghana	14.2	1.03	14	150
(10) Brazil	13.0	0.94	15	170
Other	71.6	5.19	80	1,890
TOTAL	1,379.1	100.00	1,404	38,000

*estimate
† 1 tonne = 32,151 troy ounces

IRON

Today, over 90% of the iron ore supplied to steel mills, at least of the western world, is mined from bedded iron formations of Precambrian age, mainly deposited between 2,000 and 1,600 m.y. in North America, Brazil, Sweden, European USSR, and Western Australia, and various parts of western Africa. In the earlier history of man's use of iron, there was no such predominance of Precambrian sources and industry relied more on orebodies of much younger age. For example, in England the accelerating growth of the iron and steel industry in the 18th and 19th centuries depended heavily on bodies of high-grade haematite near the west coast of Cumbria that were concentrated near the base of lower Carboniferous strata (about 350 m.y.). Later, as these mines became depleted in the late 19th century, much lower-grade bedded ores of Jurassic age (180 to 135 m.y.) were used.

The USSR has very large iron ore resources that are distributed widely both geographically and in geological age. The major production is currently from Precambrian formations near the western border.

Country	Production 1978		Production 1979*	1979 Reserves*
	Thousand Tonnes	% of World	Thousand Tonnes	Million Tonnes
(1) USSR	240,800	28.44	244,000	110,700
(2) Brazil	85,000	10.04	87,400	27,200
(3) Australia	83,189	9.83	85,400	17,800
(4) USA	82,784	9.78	85,300	25,400
(5) China (PRC)	70,000	8.27	70,000	6,100
(6) Canada	41,751	4.93	53,850	36,600
(7) India	38,155	4.51	40,630	9,100
(8) France	33,458	3.95	32,500	4,000
(9) South Africa	24,206	2.86	25,000	3,000
(10) Sweden	21,486	2.54	24,400	3,400
(11) Liberia	18,800	2.22	20,320	1,400
(12) Venezuela	13,600	1.61	13,200	2,600
Other	93,363	11.03	98,000	18,700
TOTAL	846,592	100.00	880,000	266,000

*estimate

LEAD

The largest proportion of lead production is from marine sedimentary formations ranging in age from 550 to 330 m.y. The lead is commonly accompanied by zinc and often carries economically significant silver. Most of US production is from this class. Broken Hill, Australia, the world's largest individual concentration of lead (with zinc and silver) is in much folded and altered Precambrian sediments. The largest Canadian producer, the Sullivan mine, is also of Precambrian age. A smaller proportion of world production comes from deposits of younger age than 330 m.y.

Country	Production 1978		Production 1979*	1979 Reserves*
	Tonnes	% of World	Tonnes	Thousand Tonnes
(1) USA	541,000	15.69	510,000	42,000
(2) USSR	520,000	15.09	520,000	25,000
(3) Australia	387,000	11.23	410,000	22,000
(4) Canada	366,000	10.62	340,000	20,000
(5) Peru	183,000	5.31	180,000	4,000
(6) Mexico	164,000	4.76	170,000	5,000
(7) Yugoslavia	110,000	3.19	130,000	5,000
(8) China (PRC)	100,000	2.90	100,000	7,000
(9) Morocco	91,000	2.64	92,000	4,000
(10) Sweden	82,000	2.38	85,000	3,000
(11) Spain	72,000	2.09	75,000	3,000
(12) Japan	57,000	1.65	60,000	4,000
(13) Eire	47,000	1.36	70,000	1,500
(14) South Africa	38,000	1.10	40,000	6,600
Other	689,000(1)	19.99	638,000(1)	12,900
TOTAL	3,447,000	100.00	3,420,000	165,000

*estimate
(1) Includes substantial production from Korea (DPR), Bulgaria, Poland.

MANGANESE

The largest proportion of manganese production is from Precambrian bedded deposits, commonly in the same areas as iron formations, but younger sedimentary formations also provide significant supplies. Although manganese is not by any means a rare metal, operations that are commercially competitive are heavily concentrated in six countries; (in order of production) USSR, South Africa, Gabon, Australia, Brazil and India.

When the "mining" of ocean nodules becomes economically feasible, with nickel and copper the main objectives, this will be a major source of manganese since it is the most abundant constituent.

	Production 1978		Production 1979*	1979 Reserves*
Country	Thousand Tonnes	% of World	Thousand Tonnes	Thousand Tonnes
(1) USSR	9,056	41.22	9,400	3,000,000
(2) South Africa	4,316	19.64	4,500	2,000,000
(3) Gabon	1,710	7.78	1,800	150,000
(4) Brazil	1,633	7.43	1,090	90,000
(5) India	1,566	7.13	1,630	60,000
(6) Australia	1,290	5.87	1,270	300,000
(7) China (PRC)	1,000	4.55	1,000	150,000
(8) Mexico	523	2.38	530	20,000
(9) Ghana	317	1.44	320	20,000
(10) Morocco	126	0.57	130	50,000
(11) Japan	107	0.49	100	10,000
Other	326	1.48	230	150,000
TOTAL	21,970	100.00	22,000	6,000,000[1]

* estimate

(1) Does not include ocean floor manganese nodules.

MOLYBDENUM

The demand for molybdenum for use in alloys in the steel industry has shown an annual increase of between 6 and 7%, more than that of any other base metal over recent (to 1979) years. The increase, in spite of depressed steel consumption, is partly due to the use of molybdenum in alloys for the manufacture of large diameter pipe for natural gas transmission.

By far the largest proportion of world production is from porphyry-type ore bodies, partly operations in which molybdenum is the principal product and slightly less from those where it is a by-product from other metal operations, especially copper. Since porphyry-type ore bodies are rather selectively distributed throughout the world and are particularly well-developed in the Cordilleras of North and South America, it is not surprising that over 86% of current world production comes from three countries—USA., Canada and Chile. Both USSR and China (PRC) are net importers of the metal.

	Production 1978		Production 1979*	1979 Reserves*
Country	Tonnes	% of World	Tonnes	Thousand Tonnes
(1) USA	59,802	59.88	64,000	4,400
(2) Canada	14,068	14.09	9,000	1,500
(3) Chile	13,196	13.21	14,000	1,400
(4) USSR	9,900	9.91	10,000	400
(5) China (PRC)	1,500	1.50	1,500	100
(6) Peru	729	0.73	740	400
(7) Republic of Korea	366	0.37	380	40
(8) Bulgaria	150	0.15	150	10
(9) Japan	126	0.13	130	20
(10) Philippines	24	0.02	25	60
(11) Mexico	11	0.01	15	200
TOTAL	99,872	100.00	100,000	9,800

* estimate

NICKEL

Nickel has a genetic association with mafic or ultra-mafic igneous rocks and it is probable that this metal forms a substantial proportion of the composition of the Earth's core. From the commercial standpoint, productive deposits at or near surface are of two distinct types:

(1) Sulphide concentrations associated with mafic or ultra-mafic intrusions or lava flows constitute about 20% of known reserves. Those deposits in or close to intrusions are commonly accompanied by copper and may also carry platinum group metals and gold and silver. Sudbury, Canada, is a major example and, at one time, produced a large proportion of the world supplies. The nickel-copper deposits of Norilsk, USSR, are another important example and carry an unusually high content of platinum metal. On the other hand, the nickel in ultramafic flows, such as the Western Australian deposits, has little accompanying copper or platinum metals.

(2) Nickel laterites (accounting for about 80% of known reserves) result from the weathering of ultra-mafic rocks with a low, but significant, nickel content which has been dissolved, re-circulated and concentrated by surface waters. This action takes place especially in tropical or semi-tropical climates and, consequently, those formed at the present surface are in near-equatorial latitudes, e.g. New Caledonia (the classic example), the Philippines, Indonesia, Cuba, the Dominican Republic and Colombia. Laterites formed in past geological times may be found in any latitudes but are usually covered by younger formations and are more difficult to mine.

In comparing nickel reserves in sulphides with those in laterites, it should be noted that it is much easier and cheaper to extend and establish laterite reserves, usually at or near surface, than sulphides which, as at Sudbury or Norilsk, may involve exploration at over 1,500 metres

continued overleaf

depth. Thus, the disparity between the two types of reserves is probably not as great as appears.

Ocean nodules may be a future source of nickel since this is one of the four most abundant metals contained in these concentrations.

Country	Production 1978		Production 1979*	1979 Reserves*
	Tonnes	% of World	Tonnes	Thousand Tonnes
(1) USSR	148,000	22.61	150,000	2,700
(2) Canada	127,451	19.47	120,000	7,800
(3) Australia	80,950	12.37	85,000	5,100
(4) New Caledonia	66,100	10.10	77,100	13,600
(5) Cuba	37,000	5.65	37,200	3,100
(6) Indonesia	31,914	4.88	32,000	7,100
(7) Philippines	31,046	4.74	32,000	5,200
(8) South Africa	22,500	3.44	23,000	5,700
(9) Greece	21,064	3.22	22,000	2,900
(10) Botswana	16,049	2.45	18,000	400
(11) Dominican Republic	14,300	2.19	15,000	1,000
(12) USA	12,255	1.87	12,700	2,500
(13) Zimbabwe	11,000	1.68	11,000	500
(14) China (PRC)	10,000	1.53	10,000	6,400
(15) Albania	8,000	1.22	8,000	100
Other	16,830	2.57	24,000	3,400
TOTAL	654,459	100.00	677,000	67,500

* estimate

PLATINUM GROUP METALS

This group, of which platinum and palladium are the most widely used but which includes iridium, osmium, rhodium and ruthenium, is found in mineralisation associated with mafic igneous rocks and, consequently, tends to occur with nickel-copper sulphides. They are produced both from the latter operations as by-products and also from operations where they are the principal objective. Important examples of the former are the nickel-copper mines of Norilsk, USSR, and Sudbury, Canada. The outstanding example of mineralisation mined primarily for platinum metals is found in the lower part of the layered Bushveld Complex in the Republic of South Africa. Here, a most remarkable layer, the Merensky Reef, with an average thickness of less than a metre, can be traced horizontally, with breaks, for an aggregate of 230 km. Another horizon 150 metres below has similar dimensions and below that again there is a much thicker but lower-grade layer. Together, these persistent layers contain potential reserves (estimated with some confidence) greater than the total reported reserves in the rest of the world, including USSR which has the second largest reserves and production.

Recent exploration in the layered Stillwater Complex, found in Montana, USA, has resulted in the definition of a zone of platinum metal mineralisation.

Country	Production 1978		Production 1979*	1979 Reserves*
	Tonnes †	% of World	Tonnes	Tonnes
(1) USSR	94.9	48.03	99.5	6,200
(2) South Africa	91.8	46.46	99.5	18,000
(3) Canada	8.7	4.40	7.8	280
(4) Colombia	0.4	0.20	0.5	30
(5) USA	0.2	0.10	0.3	30
Other	1.6	0.81	1.5	30
TOTAL	197.6	100.00	209.1	24,570

* estimate
† 1 tonne = 32,151 troy ounces.

SILVER

Today, over 90% of silver production is as by-product from base metal operations, especially lead, copper and zinc. Mexico is the only country where major production is now from veins that are mined primarily for silver. Cobalt, Ontario, Canada was at one time the largest world producer of silver and here silver in veins was the principal objective.

Country	Production 1978[1]		Production[1] 1979*	1979 Reserves*
	Tonnes †	% of World	Tonnes	Tonnes
(1) Mexico	1,579	14.90	1,800	26,400
(2) USSR	1,431	13.50	1,480	40,000
(3) Canada	1,280	12.07	1,240	22,000
(4) USA	1,225	11.56	1,300	47,000
(5) Peru	1,152	10.87	1,250	19,000
(6) Australia	773	7.29	800	25,000
(7) Poland	591	5.57	700	20,000
(8) Japan	303	2.85	320	1,500
(9) Chile	255	2.41	200	n.a.
(10) Bolivia	233	2.20	250	2,000
(11) Sweden	180	1.70	200	3,900
(12) Yugoslavia	159	1.50	160	1,500
(13) South Africa	96	0.91	100	7,600
(14) Spain	87	0.82	90	n.a.
(15) Honduras	97	0.92	80	800
Other	1,160	10.94	1,030	n.a.
TOTAL	10,601	100.00	11,000	250,000

* estimate
† 1 tonne = 32,151 troy ounces
(1) Includes silver recovered as a by-product of base metal ores.
Note: The above figures of reserves were based on much lower prices than those prevailing in 1980. If a price of at least $25 US per troy ounce (about $760 US per kilogram) is maintained global reserves will be markedly increased but not nearly in the same ratio as the price because of the large proportion of by-product sources.
n.a. Not available.

TIN

Tin is the most selective, geographically, of the base metals. Over 55% of known world reserves and 56% of 1979 production are from the area that includes Thailand, Malaysia, western Indonesia and southwestern China, an area about 0.45% of the surface of the globe. The mineralisation is associated with granites intruded in three main periods between 300 and 48 m.y. or in alluvial deposits derived from this mineralisation. An additional 10% of known world reserves is in Bolivia in deposits associated with intrusions dominantly of 210 to 160 m.y. age, but some much younger. Production from Precambrian sources is insignificant at present.

Production and reserves have been growing in eastern USSR. Brazil has a relatively small production, but considerable potential.

Country	Production 1978 Tonnes	Production 1978 % of World	Production 1979* Tonnes	1979 Reserves* Thousand Tonnes
(1) Malaysia	62,650	25.64	62,000	1,200
(2) USSR	33,000	13.51	34,000	1,000
(3) Bolivia	30,881	12.64	30,000	980
(4) Thailand	30,186	12.36	30,000	1,200
(5) Indonesia	24,064	9.85	24,000	1,550
(6) China (PRC)	20,000	8.19	22,000	1,500
(7) Australia	11,581	4.74	12,000	330
(8) Brazil	8,500	3.48	9,000	400
(9) Zaire	3,450	1.41	3,400	200
(10) South Africa	2,887	1.18	2,900	150
(11) United Kingdom	2,831	1.16	2,600	260
(12) Nigeria	2,751	1.13	3,000	280
(13) Peru	1,744	0.71	1,750	160
(14) Rwanda	1,700	0.70	1,700	160
(15) Germany (DDR)	1,600	0.65	1,600	150
Other	6,488	2.66	7,050	480
TOTAL	244,313	100.00	247,000	10,000

*estimate

TITANIUM

The two principal minerals that are the source of the titanium in alloys (especially for aircraft and space use) and for pigment and other industrial uses, are rutile (TiO_2) and ilmenite [$(FeTi)_2O_3$]. Both are resistant minerals easily concentrated by current or wave action into "heavy sands". Consequently, the mining or dredging of such heavy sands, either on land or off-shore, accounts for a major proportion of world production.

Although rutile rarely occurs in mineable primary form in rock, ilmenite concentrations are found in basic rocks, especially anorthosite. The most important production from such primary sources is from southern Norway and from the province of Quebec, north of Havre St. Pierre on the St. Lawrence River. Both deposits are in intrusive anorthosite of late Precambrian age, about 900 m.y. The Norwegian operation is a combined iron and titanium producer, the ore carrying an average of 18% titanium oxide. The Quebec occurrence is mined primarily for titanium and the ilmenite is converted to "titaniferous slag" which is the product sold.

Country	1978 Production[1] Thousand Tonnes of Concentrate	1978 Production[1] Thousand Tonnes of Contained TiO_2	% of World	Production 1979* Thousand Tonnes of Concentrate	1979 Reserves* Thousand Tonnes of Concentrate
(1) Australia	1,525	894	31.46	1,420	56,000
(2) Canada	850[2]	595	17.53	525	166,000
(3) Norway	767	391	15.82	770	140,000
(4) USA	535[3]	273	11.04	500	51,000
(5) USSR	440	238	9.08	450	50,000
(6) Malaysia	187	95	3.86	190	25,000
(7) India	151	80	3.11	155	107,000
(8) Finland	132	67	2.72	135	20,000
(9) S. Africa	108	94	2.23	340	34,000
(10) China (PRC)	95	51	1.96	100	20,000
Other	58	31	1.20	55	21,000
TOTAL	4,848	2,809	100.00	4,640	690,000

*estimate

(1) The figures of 'concentrate' are principally from U.S. Bureau of Mines publications. Those of 'tonnes of contained TiO_2' are calculated from the above. Since, however, the proportion of ilmenite and rutile is not always published and the titanium dioxide content varies between 51% in the former and 95% in the latter, the figures should be regarded as approximate.

(2) In the form of titaniferous slag containing about 70% TiO_2.

(3) Does not include a small tonnage of rutile.

TUNGSTEN

Although the metal tends to be associated with tin in natural occurrence, the dominant production is not from the two most important tin-producing areas of the world. The largest production (and probably the greatest potential) is in southeast China, where the containing mineralisation (with subordinate tin) is associated with granites intruded between 150 and 120 m.y.

In North America, a single mine in northwest Canada supplied 4.8% of world production in 1978 and a large deposit in Nevada is under development.

Tungsten occurs in small quantities in brines and the possibility of commercial recovery from this source is under investigation.

Country	Production 1978 Tonnes	Production 1978 % of World	Production 1979* Tonnes	1979 Reserves* Tonnes
(1) China (PRC)	9,000	19.66	9,000	1,000,000
(2) USSR	8,500	18.57	9,000	215,000
(3) Australia	3,379	7.38	3,900	110,000
(4) Bolivia	3,170	6.92	3,200	40,000
(5) USA	3,130	6.84	3,000	125,000
(6) Thailand	2,942	6.43	1,800	20,000
(7) Republic of Korea	2,681	5.86	2,700	55,000
(8) Canada	2,279	4.98	2,700	245,000
(9) Korea (DPR)	2,150	4.70	2,200	45,000
(10) Brazil	1,179	2.58	1,200	20,000
(11) Austria	1,179	2.58	1,400	20,000
(12) Portugal	1,104	2.41	1,200	25,000
(13) Japan	761	1.66	800	20,000
(14) Burma	670	1.46	670	30,000
Other	3,657	8.00	3,700	530,000
TOTAL	45,781	100.00	46,500	2,500,000

*estimate

ZINC

This metal is derived mainly from two distinct classes of deposits. The first, which provides by far the largest proportion, is in marine sedimentary formations, especially limestones between the ages of 550 and 330 m.y., and commonly, but not exclusively, associated with lead. Some important production is from Precambrian sediments in Australia and Canada. The second class, in which zinc is typically accompanied by copper, is from massive sulphide bodies associated with acid volcanic activity of various ages, but favouring Precambrian.

Country	Production 1978		Production 1979*	1979 Reserves*
	Tonnes	% of World	Tonnes	Thousand Tonnes
(1) Canada	1,244,000	19.89	1,250,000	62,000
(2) USSR	850,000	13.59	900,000	12,000
(3) Peru	458,000	7.32	480,000	7,000
(4) Australia	441,000	7.05	490,000	24,000
(5) USA	337,000	5.39	280,000	48,000
(6) Japan	275,000	4.40	270,000	4,500
(7) Mexico	271,000	4.33	250,000	3,000
(8) Eire	173,000	2.77	206,000	7,000
(9) Sweden	163,000	2.61	160,000	3,000
(10) Spain	141,000	2.25	140,000	10,000
(11) China (PRC)	125,000	2.00	125,000	1,800
(12) Germany (FDR)	117,000	1.87	120,000	1,800
(13) South Africa	110,000	1.76	120,000	15,400[2]
(14) Yugoslavia	86,000	1.38	100,000	1,000
(15) Greenland	83,000	1.33	88,000	900
(16) Zaire	72,000	1.15	75,000	1,800
(17) Italy	71,000	1.14	75,000	3,600
Other	1,237,000[1]	19.78	1,216,000	33,200
TOTAL	6,254,000	100.00	6,345,000	240,000

* estimate

(1) Includes substantial production from Poland, Korea (DPR), Bulgaria.

(2) Includes Namibia.

ENERGY MINERALS

COAL

Coal did not begin to be formed until about 340 m.y. when land plants became sufficiently abundant to accumulate in layers interbedded with clays and sands. In North America, coal formed in two main periods, the first from about 310 to 270 m.y. and the second from about 100 to 50 m.y. The first corresponds roughly to the period of the principal generation of coal in Europe which gave rise to the name "Carboniferous Period". The coal reserves of the USSR and the still more extensive resources of China range all the way from 320 to 30 m.y.

Carbonaceous deposits in sediments younger than 30 m.y. are usually in the form of lignites or brown coal.

Country	Production 1978		Production 1979*	1979 Reserves*
	Million Tonnes	% of World	Million Tonnes	Million Tonnes
(1) China (PRC)	618.0	23.50	670.0	98,883
(2) USA	565.5	21.50	600.0	113,230
(3) USSR	555.6	21.13	580.0	82,900
(4) Poland	192.7	7.33	200.0	20,000
(5) United Kingdom	119.9	4.56	125.0	45,000
(6) India	101.3	3.85	105.0	10,683
(7) South Africa	90.6	3.45	100.0	26,903
(8) Germany (FDR)	83.5	3.18	90.0	23,919
(9) Australia	75.0	2.85	80.0	18,128
(10) Korea (DPR)	48.0	1.83	50.0	300
(11) Czechoslovakia	28.3	1.08	30.0	2,493
(12) Canada	24.4	0.93	30.0	3,663
Other	127.0	4.83	140.0	14,468
TOTAL	2,629.8	100.00	2,800.0	460,570

* estimate

Note: There is substantial production of brown coal from (1) Germany (DDR), (2) USSR, (3) Germany (FDR), (4) Yugoslavia.

NATURAL GAS

Natural gas occurs in the same general areas and formations as oil. However, since gas results from generally deeper burial (and hence higher pressure and temperature) than does oil, it tends to favour somewhat older formations. Thus in the "Giant Gas Fields" reserves of the world ("Giant" being defined as one with an ultimate recoverable reserve of 3 trillion cubic feet) 16.3% is in sediments older than 225 m.y., 62.2% in those dating 225 to 65 m.y. and 21.3% in beds younger than 65 m.y.

The USSR has a much higher proportion of reported world gas reserves (39%) than it has of oil (11.4%). However, the country with the highest production of natural gas is the USA, although the largest reserves are in the USSR.

Country	Production 1978		Production 1979[1]	1980 Reserves
	10^9 cu metres	% of World	10^9 cu metres	10^9 cu metres
(1) USA	564	37.00	567	5,493
(2) USSR	372	24.38	404	25,485
(3) Canada	89	5.82	104	2,421
(4) China (PRC)	66	4.32	79	708
(5) Iran	49	3.24	26	13,875
(6) Netherlands	46	3.02	76	1,685
(7) Romania	34	2.25	34	n.a.
(8) Mexico	25	1.65	27	1,671
(9) Germany (FDR)	18	1.19	24	180
(10) Nigeria	17	1.09	10	1,172
(11) Indonesia	16	1.05	23	680
(12) Italy	15	1.00	15	99
(13) Venezuela	11	0.74	12	1,212
(14) Libya	10	0.68	5	680
(15) Norway	10	0.65	12	665
(16) Algeria	9	0.61	9	3,738
Other	173	11.31	190	13,102
TOTAL	1,525	100.00	1,617	72,866

(1) Based on 11 months of 1979.

n.a. Not available.

OIL

Petroleum has been generated in marine sediments at least as old as 550 m.y. and the earliest commercial production (in Ontario, Canada, in 1857) was from Silurian and Devonian beds, about 450 to 345 m.y. old. However, by far the largest proportion of world supplies today is from beds younger than 225 m.y. For example, of the "Giant Oil Field" reserves of the world ("Giant" being defined as over 70 million tonnes), only about 5.4% is in sediments older than 225 m.y., 68.8% are in those dating 225-65 m.y. and 24.5% are in beds younger than 65 m.y. The Persian/Arabian Gulf oil fields, which contain over 65% of the known world reserves, are nearly all in sediments younger than 225 m.y., as are those of the North Sea, Nigeria and Venezuela. The producing beds in North America have a higher *average* age than those of the fields just mentioned.

Country	Production 1978 10³ bbls/day	% of World	Production 1979* 10³ bbls/day	Reserves 1980* 10⁶ bbls
(1) USSR	11,400	19.00	11,670	67,000
(2) United States	8,660	14.44	8,650	26,500
(3) Saudi Arabia	7,800	13.00	9,250	163,350
(4) Iran	5,250	8.75	2,900	58,000
(5) Iraq	2,500	4.17	3,370	31,000
(6) Venezuela	2,150	3.58	2,330	17,870
(7) Libya	2,050	3.42	2,050	23,500
(8) China (PRC)	2,000	3.33	2,100	20,000
(9) Kuwait	1,900	3.17	2,210	65,400
(10) Nigeria	1,800	3.00	2,370	17,400
(11) Indonesia	1,650	2.75	1,600	9,600
(12) Abu Dhabi	1,450	2.42	1,450	28,000
(13) Canada	1,300	2.17	1,480	6,800
(14) Mexico	1,270	2.12	1,800	50,000
(15) Algeria	1,260	2.10	1,240	8,440
(16) United Kingdom	1,100	1.83	1,570	15,400
(17) Egypt	490	0.82	500	3,100
(18) Qatar	480	0.80	480	3,760
Other	5,480	9.13	5,880	45,253
TOTAL	59,990	100.00	62,590	660,373

*estimate

URANIUM

Uranium is not a very scarce metal in the Earth's crust but commercial grade concentrations occur mainly in rocks of pre-1,000 m.y. and post-230 m.y. age with rather limited resources between these ages.

The older group includes several types of deposits including the conglomerate deposits of the Witwatersrand, South Africa and Elliot Lake, Canada (about 2,600 and 2,200 m.y. age respectively) and deposits associated with unconformities dating between 1,700 and 1,200 m.y. In the latter class, there is current production, and much more additional reserves not yet in production, in Canada, and extensive reserves under development in Australia. A third type, associated with Precambrian felsic intrusives, has resulted in production in Canada and Australia but the largest example is the Rössing Mine in Namibia.

The younger group, where uranium is concentrated in relatively unaltered sediments, accounts for nearly all production up to 1979 in USA which at that date provided about 35% of western world production.

Although USSR in 1978 appeared to be arranging to import uranium fuel from overseas, it is thought that, in the long run, adequate resources of this metal will be developed.

In the People's Republic of China, uranium resources are known in several areas but the low per capita consumption of energy and the ready availability of low-cost coal makes it unlikely that nuclear energy will be widely developed in the near future.

Country	Production 1978 Tonnes U₃O₈	% of World	Production 1979* Tonnes U₃O₈	1979 Reserves* Tonnes U₃O₈
(1) USA	16,783	35.86	23,900	620,000
(2) Canada	8,005	17.11	8,200	620,000
(3) USSR	7,000	14.96	8,000	500,000
(4) South Africa	4,659	9.96	6,900	515,000
(5) Namibia	3,719	7.95	4,500	90,000
(6) Niger	2,857	6.10	4,550	160,000
(7) France	1,608	3.43	4,550	65,000
(8) Gabon	616	1.31	1,415	30,000
(9) Australia	608	1.29	600	432,000
Other	952	2.08		18,000
TOTAL	46,807	100.00		3,050,000

*estimate

GEOTHERMAL ENERGY

The internal heat of the Earth is a natural resource which has been increasingly considered an exploitable alternative source of energy, with water (in various states) being the transporting medium.

There are two main sources of geothermal energy.

(1) Hot rock, into which cooler water, probably of surface origin, seeps and is converted into steam. It is subsequently tapped for use, at the surface, in the production of electrical energy. This source of geothermal energy is referred to as "dry steam".

(2) Hot water or steam may be trapped under pressure in porous formations. When tapped by drilling, the water "flashes" into steam as it reaches lower pressures near the surface. This potential source of geothermal energy is known as "wet steam".

The earliest example of the production and use of geothermal energy was near Larderello in northwest Italy in 1904. This is the "dry steam" type as is the Big Geyser Field which was the first developed in the US in the mid-1950's. The Wairakei Field, in North Island, New Zealand, was the second geothermal occurrence to be developed commercially and is the "wet steam" type, as are the geothermal fields in Mexico and most of those in the United States other than the Big Geyser Field.

Geothermal power production is expected to grow rapidly in the next few years in many countries. Most geothermal sources are either in the vicinity of geologically young volcanic activity or are near young faults which penetrate rocks with significantly high temperatures at depth.

The following list shows the approximate power generated from geothermal sources at the end of 1978, broken down by countries.

Country	Capacity (megawatts)	Comment
USA	502	—Up to 1979 production solely from big geysers "dry steam" occurrences.
Italy	380	—Larderello "dry steam".
New Zealand	200	—Wairakei "wet steam".
Japan	170	—about 13% "dry", remainder "wet" steam.
Mexico	75	—Cerro Prieto area "wet steam".

SUGGESTED READING

It is clearly impossible in a publication of this type to include any comprehensive reference list or to do more than suggest a few publications that will give reliable additional information on the respective aspects.

Introductory Geology

For anyone who has little or no geological background, one of the most concise, informative and well-illustrated introductions to geology is produced by the Institute of Geological Sciences at the Geological Museum in London, England. Entitled *The Story of the Earth*, it contains, in only 36 pages, an admirable amount of information and is modestly priced.

In somewhat more detail but still in layman's language, while based on current concepts, is J. R. Janes' (1979) *Geology and the New Global Tectonics*, published by MacMillan of Canada. This is suitable for the student of geology in a secondary school or first year university.

A good review and explanation of plate tectonics, in layman's language and with entertaining illustrations, is Anthony Hallam's *A Revolution in the Earth Sciences* (1973) published by Oxford University Press.

General Geology

There are many textbooks written for undergraduate use. One of the best known and most respected in the English language is the latest edition of *Principles of Physical Geology* by Arthur Holmes, revised (1978) by Doris Holmes and published by Thos. Nelson.

Other good textbooks covering the general field of physical geology are:-

Geology by W. C. Putnam, revised by Ann Bassett (1971), published by Oxford University Press.

Physical Geology, 5th edition, by L. D. Leet, S. Judson and M. E. Kauffman published by Prentice Hall in 1978.

Dealing especially with modern concepts is *The Way the Earth Works* by Peter J. Wyllie (1977), published by John Wiley & Sons Inc.

Geology of Mineral Deposits

A good introduction to the economic geology field is *The Earth's Physical Resources*, Block 3, Mineral Deposits, by Open University Course Team, Open University Press, England.

More advanced, but concisely written, is Brian J. Skinner's *Earth Resources*, 2nd Edition (1976), published by Prentice Hall Inc.

Probably the most comprehensive textbook on metallic deposits, on a global scale, is Pierre Routhier's *Les Gisements Métallifères—Géologie et Principes de Récherche* (1963), in two volumes, published by Masson et Cie.

A valuable reference on ore deposits in the Americas is Geological Society of America Memoir 131 *Annotated Bibliography of Mineral Deposits in the Western Hemisphere* (1972) by J. D. Ridge.

Unique is the *Atlas of Economic Mineral Deposits* by Colin Dixon (1979), published by Chapman and Hall. In summarised text, supplemented by geological plans and sections, the origin and controlling features of 40 important mineral deposits, metallic and non-metallic, are presented in a form suitable for the professional or the graduate student.

In petroleum geology, a very good review of the genesis and localisation of oil and gas on a global basis is *Petroleum and Global Tectonics* by A. G. Fischer and Sheldon Judson (1975), published by Princeton University Press.

Global Mineral Production and Reserves

A book entitled *World Mineral Supplies - Developments in Economic Geology No. 3* by G. J. S. Govett and M. H. Govett (1976), published by Elsevier Scientific Publishing Co., is recommended for general coverage of the subject.

To update the figures quoted in this Atlas regarding mineral production and reserves, the reader is referred to the following publications:-

(1) *Mineral Commodity Summaries*, published yearly about February by the Bureau of Mines, US Department of the Interior. This gives an up-to-date summary for the previous year of 90 (solid) mineral commodities including production and reserve figures broken down by nations.

(2) *Mineral Trade Notes*, also published by the Bureau of Mines, US Department of the Interior, is released several times a year, each number covering selected minerals and metals and giving more detailed production figures than the previous reference.

(3) *Mining Annual Review*, published in June of each year by Mining Journal Ltd, 15 Wilson Street, London, covers world production in the previous calendar year, and known reserves of each important metal or solid mineral broken down by producing countries. In a separate section each country is listed with a review of its mineral production, development and exploration.

(4) *World Mining*, published monthly in San Francisco, by Miller Freeman Publications, produces an annual review about July of each year. Although less comprehensive in statistics, the mineral developments on a global scale, and new techniques and equipment, are reviewed.

(5) *World Coal*, published monthly in London, by Miller Freeman Publications gives a statistical summary about November of each year. This gives figures for production in the previous calendar year of the various ranks of coal, broken down by producing countries, and known or estimated reserves and resources.

(6) *Oil and Gas Journal*, annual worldwide issue, published by Petroleum Publishing, usually in the last weekly number of the year. On the basis of production figures for the first six months of the year it gives up-to-date estimates for the current year of world oil and gas output, by countries, as well as available figures on reserves.

Below are listed, under headings corresponding to the individual map sheets, addresses where maps and reports can be obtained that give more detailed information on geology and mineral deposits of the respective countries.

AMERICAS SHEET

Servicio Geológico Nacional
Secretaria de Estado de Recursos Naturales
Avenida Santa Fe 1548
Buenos Aires, Argentina

Bellairs Research Institute
St James, Barbados

Servicio Geológico de Bolivia
Ministerio de Minas y Petroleo
Federico Zuazo, Esq. Reyes Ortiz
Casilla de Correo 2729
La Paz, Bolivia

Departmento Naçional de Produçao
   Mineral (DNPM)
Setor de Autarquia Norte
Esplanada dos Ministerios Quadra 1-
   Projeçáo F
Brasília, D.F., Brazil

Geological Survey of Canada
Department of Energy, Mines and Resources
601 Booth Street
Ottawa, Ontario, K1A OE8, Canada

Dirección de Geologia y Minas
Corporación Nacional del Cobre de Chile
Huérfanos 1189, 4° Piso, Santiago, Chile

Instituto Nacional de Investigaciones
   Geológico-Mineras
Carrera 30, No. 51-59
Apartado Nacional 2504
Bogotá, D.E., Colombia

Departamento de Geologia
Instituto Costarricense de Electricidad
Apartado 10032
San José. Costa Rica

Instituto de Geología
Academia de Ciencias de Cuba
Ave. Van-Troi no. 17203
Rancho Boyeros-Apartado Postal 10
La Habana, Cuba

Department of Trade and Industry
Roseau, Dominica

Dirección de Mineria
Secretaria de Estado de Fomento
Santo Domingo, Dominican Republic

Dirección General de Geología y Minas
Ministerio de Recursos, Naturales y
   Energéticos
Carrion # 1016 y Páez
Quito, Ecuador

Centro de Investigaciones Geotécnicas
Ministerio de Obras Públicas
Avenida Peralta, final, costado Oriente
   Talleres El Coro
San Salvador, El Salvador

Grønlands Geologiske Undersøgelse
Ostervoldgrade 5—7, Tr. KL
DK-1350 København, K. Denmark

Bureau des Recherches Géologiques et
   Minières, B.P. 42
Cayenne, French Guiana

Geological Surveys and Mines Department
P.O. Box 1028
Georgetown, Guyana

Service Géologique
   Department of Agriculture, Natural
   Resources, and Rural Development
Damiens près Port-au-Prince, Haiti

Dirección General de Minas e
   Hidrocarburos
Boulevard a la Colonia Kennedy
Tegucigalpa, D.C. Honduras

Geological Survey Department
Hope Gardens
Kingston 6, Jamaica

Arrondissement Minéralogique de la
   Guyane, B.P. 458
Fort-de-France, Martinique

Instituto de Geologia
Universidad Nacional Autonoma de
   México (UNAM)
Ciudad Universitaria
México 20, D.F., Mexico

Servicio Geológico Nacional
Ministerio de Economia
Apartado Postal 1347
Managua, D.N., Nicaragua

Administración de Recursos Minerales
Apartado Postal 9658, Zona 4
Panama City, Panama

Dirección de Recursos Minerales
Ministerio de Obras Publicas and
   Comunicaciones
Calle Alberdi y Oliva
Asunción, Paraguay

Instituto de Geología y Minería
   (INGEOMIN)
Paz Soldan 225
San Isidro, Apartado 889
Lima, Peru

Geologisch Mijnbouwkundige Dienst
Klein Wasserstraat 1 (2-6)
Paramaribo, Surinam

US Geological Survey
National Center
12201 Sunrise Valley Drive
Reston, Virginia 22092
USA

Instituto Geológico del Uruguay
Calle J. Herrera y Obes 1239
Montevideo, Uruguay

SOURCES OF ADDITIONAL INFORMATION

Dirección General de Minas y Geologia
Ministerio de Energía y Minas
Torre Norte, Piso 27
Centro Simón Bolívar
Caracas, Venezuela

NORTHERN EUROPE SHEET

Service Géologique de Belgique
13 rue Jenner, Parc Léopold
Bruxelles 4, Belgium

Institute of Geological Sciences
Exhibition Road, South Kensington
London, SW7 2DE, Britain

Danmarks Geologiske Undersøgelse
Thoravej 31
DK 2400 København N.V. Denmark

Geologinen Tutkimuslaitos
02150 Espoo 15, Finland

Zentrales Geologisches Institut
Invalidenstrasse 44
104 Berlin, Germany (DDR)

Bundesanstalt für Geowissenschaften
und Röhstoffe
Postfach 510153, Stilleweg 2
3000 Hannover 51, Germany (FDR)

Geological Survey of Ireland
14 Hume Street
Dublin 2, Ireland

Rijks Geologische Dienst
Spaarne 17, P.O. Box 157
Haarlem, Netherlands

Norges Geologiske Undersøkelse
P.B. 3006 Ostmarkneset
Leiv Erikssons Vei 39
7001 Trondheim, Norway

Instytut Geologiczny
ul. Rakowiecka 4
00-975 Warszawa, Poland

Sveriges Geologiska Undersökning (SGU)
[The Geological Survey of Sweden]
S-10405 Stockholm 50, Sweden

Geologic Institute
Akademiya Nauk SSSR
Pyzhevsky per. 7
Moscow G-17, USSR

MEDITERRANEAN EUROPE SHEET

Ministry of Industry and Mining
Tirane, Albania

Geologische Bundesanstalt
Rasumovskygasse 23
A-1031 Vienna, Austria

Service Géologique de Belgique
13 rue Jenner, Parc Léopold
Bruxelles 4, Belgium

Geologicheski Institut
Bulgarska Akademiya na Naukite
Akademik Bonchev St., Blok 2
Sofia, Bulgaria

Geological Survey Department
Ministry of Agriculture and Natural
Resources
P.O. Box 809
Nicosia, Cyprus

Ústav Geologický
[Geological Institute]
Ceskoslovenska Akadémie Ved
[Czechoslovak Academy of Sciences]
Lysolaje 6
Posta Suchodol 2
Praha 6, Czechoslovakia

Bureau des Recherches Géologiques et
Minières (BRGM)
B.P. 6009
45018 Orléans Cedex, France

Nazional Kommittee Für Geologische Wissenschaften
Otto Nuschke Strasse 22-23
DDR-108 Berlin, Germany (DDR)

Zentrales Geologisches Institut
Invalidenstrasse 44
104 Berlin, Germany (DDR)

Bundesanstalt für Geowissenschaften
und Rohstoffe
Alfred-Bentz-Haus
Postfach 510153 Stilleweg 2
3000 Hannover 51, Germany (FDR)

Institute of Geological and Mining Research
70 Messoghion Street
Athens 608, Greece

Magyar Allami Foldtani Intézet
[Hungarian State Geological Institute]
Nepstadion-ut 14, Pf. 106
Budapest 1442, Hungary

Servizio Geologico d'Italia
Ministero dell'Industria del Commercio
e dell'Artigianato
Direzione Generale della Minere
Largo di Santa Susanna 13
Roma, Italy

Serviços Geológicos de Portugal
Rua da Academia das Ciências 19-2º
Lisboa-2, Portugal

Instytut Geologiczny
ul. Rakowiecka 4
00-975 Warszawa, Poland

Institutul de Geologie si Geografie
Str. Dimitrie Racovitá No. 9
Bucharest, Romania

Instituto Geológico y Minero de España
Rios Rosas 23
Madrid, Spain

Schweizerische Geologische Kommission
Naturforschende Gesellschaft
Bernoullistrasse 32
4056 Basel, Switzerland

Ministry of Energy and Natural Resources
Bakanliklar
Ankara, Turkey

Zavod za Geoloski i Geofizička
Instrazivanja (GEOZAVOD)
Karadjordjeva 48
Belgrade, Yugoslavia

AFRICA SHEET

Service Géologique de l'Algérie
Immeuble Mauritanie
Boulevard Colonel Amirouche
Agha, Algiers, Algeria

Direcção de Servicos de Geologia e Minas
Caixo Postal 1260-C
Luanda, Angola

Direction des Mines, de la Géologie
  et des Hydrocarbures
Ministère des Travaux Publiques, des
  Transports et des Télécommunications
B.P. 249
Cotonou, Benin

Geological Survey Department
Ministry of Mineral Resources and
  Water Affairs
Private Bag 14
Lobatsi, Botswana

Financières, Bukundi
Department de Géologie et Mines
B.P. 745, Bujumbura, Burundi

Direction des Mines et de la Géologie
Ministry of Mines and Energy
B.P. 70
Yaoundé, Cameroons

Secretary of State for Mines and Energy
Bangui, Central African Republic

Direction des Mines et de la Géologie
B.P. 816
N'Djamena, Chad

Service des Mines et de la Géologie
B.P. 12
Brazzaville, Rep. of the Congo

Services de Travaux Publiques
Djibouti, Djibouti

Geological Survey and Mining Authority
Ministry of Industry
Abbasia Post Office
Cairo, Egypt

Ethiopian Geological Survey Institute
Ministry of Mines, Energy and Water
P.O. Box 486
Addis Ababa, Ethiopia

Bureau de Recherches Géologiques et
  Minières (BRGM)
B.P. 175
Libreville
Gabon

Survey Department
Banjul, The Gambia

Geological Survey of Ghana
P.O. Box M80
Accra
Ghana

Services des Mines et de la Géologie
Conakry
Guinea

Serviços de Geologia e Minas
Caixa Postal 399
Bissau, Guinea-Bissau

Geological Survey of Israel
30 Malchei Israel Street
Jerusalem, Israel

Direction des Mines et de la Géologie
Ministère des Mines
B.P. V 28
Abidjan
Ivory Coast

Mines and Geological Survey Department
P.O. Box 39
Ministry of National Economy
Amman
Jordan

Geological Survey of Kenya
Mines and Geological Department
Ministry of Natural Resources
P.O. Box 30009
Nairobi, Kenya

Kuwait Institute for Scientific Research
P.O. Box 12009 Shamiah
Al Kuwayt, Kuwait

Direction Génerale des Travaux Publiques
Ministère des Ressources Naturelles et
  Energie Hydraulique
Beirut, Lebanon

Liberian Geological Survey
Ministry of Lands and Mines
P.O. Box 9024
Monrovia, Liberia

Geological Research and Mining
  Department
Industrial Research Center
P.O. Box 3633
Tripoli, Libya

Service Géologique
Ministère des Mines et de L'Energie
B.P. 280, Armandrianomby
Tananarive, Malagasy Republic

Geological Survey Department
Ministry of Natural Resources
P.O. Box 27, Liwonde Road
Zomba, Malawi

Direction National des Mines et de la
  Géologie
B.P. 223
Koulouba, Bamako, Mali

Direction des Mines et de la Géologie
Ministère de l'Industrialisation et des Mines
B.P. 199
Nouakchott, Mauritania

Division de la Géologie
Ministère du Commerce, de l'Industrie
  des Mines, et de la Marine Marchande
Rabat, Morocco

Direcção dos Serviços de Geologia e Minas
Caixa Postal 217
Maputo, Mozambique

Geological Survey
P.O. Box 2168
Windhoek, Namibia

SOURCES OF ADDITIONAL INFORMATION

Ministère des Mines et de l'Hydraulique
B.P. 257
Niamey, Niger

Nigerian Geological Survey Department
Ministry of Mines and Power
P.M.B. 2007
Kaduna South, North State, Nigeria

Ministry of Agriculture, Fisheries,
Petroleum, and Natural Resources
P.O. Box 551
Muscat, Oman

Qatar Petroleum Producing Authority
(QPPA)
P.O. Box 70
Doha, Qatar

Geological Survey of Zimbabwe
Ministry of Mines and Lands
P.O. Box 8039, Causeway
Salisbury, Zimbabwe

Service Géologique du Rwanda
Ministère du Commerce, des Mines et de
l'Industrie
B.P.15
Ruhengera, Rwanda

Ministry of Petroleum and Mineral
Resources
Directorate General of Mineral Resources
P.O. Box 345
Jiddah, Saudi Arabia

Bureau de Recherches Géologiques et
Minières (BRGM)
7, rue Mermoz
B.P. 268
Dakar, Senegal

Institut Français Afrique Noire
BP 206
Dakar, Senegal

Geological Survey Division
Ministry of Lands, Mines, and Labor
New England, Freetown
Sierra Leone

Geological Survey Department
Ministry of Mining
P.O. Box 744
Mogadishu, Somalia

Geological Survey of South Africa
Department of Mines
Private Bag X1112
New Museum Building
233 Visagie Street
Pretoria 0001
South Africa

Geological Survey Department
Ministry of Mining and Industry
P.O. Box 410
Khartoum, Sudan

Directorate of Geological Research and
Mineral Resources
Ministry of Petroleum
Fardos Street
Damascus, Syria

Ministry of Water Development, Power,
Energy and Minerals
P.O. Box 412
Dodoma, Tanzania

Direction des Mines et de la Géologie
Ministère des Mines, Energie et Ressources
Hydrauliques
B.P. 356
Lomé, Togo

Direction des Mines et de la Géologie
Ministère de l'Economie Nationale
195 rue de la Kasbah
Tunis, Tunisia

Direction de la Géologie et des Mines
B.P. 601
Ougadougou, Upper Volta

Office of Mineral Resources
Minerals and Petroleum Authority
Ministry of Public Works
San'ā, North Yemen

Service Géologique du Zaire
Ministère des Mines
D.P. 898
44 Avenue des Huileries
Kinshasa, Zaire

Geological Survey Department
Ministry of Mines and Industry
P.O. Box RW 135
Ridgeway, Lusaka, Zambia

U.S.S.R. SHEET
All-Union Geological Scientific Research
Institute (VSEGEI)
Sredny Prospekt 72B
Leningrad 199026, USSR

Geologic Institute
Akademiya Nauk SSSR
Pyzhevsky per. 7,
Moscow G-17, USSR

Ministry of Geology of the USSR
Bolshaya Gruzinskaya, 416
Moscow D-242, USSR

SOUTHERN ASIA SHEET
Department of Mines and Geology
Ministry of Mines and Industries
Darulaman
Kabul, Afghanistan

Geological Survey
Pioneer Road
Segun Bagicha
Dacca, Bangladesh

Department of Trade and Industries
Thimphu, Bhutan

Geological Division
Ministry of Agriculture and Forestry
Rangoon, Burma

Ministry of Industry, Mines
and Mineral Prospecting
Phnom Penh, Kampuchea

Institute of Geology
Chinese Academy of Sciences
Peking, People's Republic of China

Mineral Resources Division
Private Mail Bag, G.P.O.
Government Buildings
Suva, Fiji

Geological Survey of India
27 Jawaharlal Nehru Road
Calcutta 700013, India

Geological and Mineral Survey of Iran
Ministry of Industry and Mines
P.O. Box 1964
Tehran, Iran

Department of Geological Survey and
   Mineral Investigation
Ministry of Oil
P.O. Box 986
Alwiyah, Baghdad, Iraq

Geological Survey of Japan, Ministry of
   International Trade and Industry
135 Hisamoto-cho, Takatsu-ku
Kawasaki-shi, Kanagawa-ken 213, Japan

Geology and Geography Research Institute
Academy of Sciences
Mammoon-dong
Central District
P'yongyang, Korea (DPR)

Korea Research Institute of Geoscience
   and Mineral Resources
219-5, Garibong-dong
Youngdeungpo-gu
Seoul 150-06, Rep. of Korea

Service de Géologie
Comité Central du Plan
Vientiane, Laos

Institute of Geology
Academy of Sciences
Ul, Leniadom 2
Ulan Bator, Mongolia

Department of Mines and Geology
Ministry of Industry and Commerce
Lainchaur
Kathmandu, Nepal

Geological Survey of Pakistan
P.O. Box 15
Quetta, Pakistan

Geological Survey Department
48 Sri Jinaratana Road
Colombo 2, Sri Lanka

Geological Survey of Taiwan
P.O. Box 1001
Taichung 400, Taiwan

Department of Mineral Resources
Ministry of Industry
Rama VI Road
Bangkok, Thailand

Geology Section
State Committee of Sciences
Hanoi, Vietnam

INDONESIA-AUSTRALASIA SHEET

Department of National Resources
P.O.Box 378
Canberra, A.C.T. 2600, Australia

The Government Geologist
Brunei Town, Brunei

Geological Survey of Indonesia
Jalan Diponegoro 57
Bandung, Indonesia

Federal Geological Service
Jalan Gurney
Kuala Lumpur 15-01, Malaysia

Service des Mines et de la Géologie
Rte. No. 1, B.P. 465
Noumea, New Caledonia

Geological Survey
British Residency
Port Vila, New Hebrides

New Zealand Geological Survey
Department of Scientific and Industrial
   Research, P.O. Box 30-368
Wellington, New Zealand

Geological Survey of Papua-New Guinea
Box 778
Port Moresby, Papua-New Guinea

Geological Survey Division
Bureau of Mines
P.O. Box 1596, Pedro Gill St.
Manila 2801, Philippines

Department of Geological Surveys
P.O. Box G24
Honiara, Guadalcanal, Solomon Islands

Ministry of Lands, Surveys and Natural
   Resources, P.O. Box 5
Nuku' alofa, Tonga

ARCTIC SHEET

Department of Economic Geology
National Energy Authority
Laugavegi 116
105 Reykjavik, Iceland

GENERAL REFERENCES

American Association of Petroleum
   Geologists (AAPG)
P.O. Box 979
Tulsa, Oklahoma 74101 USA
Executive Director: Fred Dix

Commission for the Geological
   Map of the World
51, Boulevard de Montmorency
75026 Paris, France
Secretary General: Frances Delany

International Union of Geological
   Sciences (IUGS)
Geological Survey of Canada,
601 Booth Street
Ottawa, Ontario
K1A 0E8, Canada
Secretary General: W. Hutchison

GLOSSARY

While the text of this Atlas has been written in layman's language, as far as possible, the use of some words that may not be familiar to the reader with limited earth science background cannot be avoided. Below is a Glossary covering 100 or so words and expressions that may need explanation.

For additional definitions the reader is referred to either the *Penguin Dictionary of Geology* by D.J.A. Whitten and J.R.V. Brooks (published by Penguin Books) or the *Dictionary of Geological Terms* (revised edition) prepared under the direction of The American Geological Institute (published by Anchor Press/Doubleday, Garden City, New York). Both are pocket-sized. The former is aided by line sketches in many of the explanations, while the latter is the more comprehensive.

Most of the definitions in this Glossary are derived from one or other of the above publications.

Alluvial   Deposited by rivers or streams, e.g. alluvial gold (placer).

Andesite   Lava of intermediate composition between basic (mafic) and acidic (felsic).

Anorthosite   A plutonic rock composed mainly of plagioclase feldspar with a relatively high calcium content.

Anticline   A fold that is (at least originally) convex upward. Overfolding may result in the axis being horizontal and further folding may result in the fold being turned upside down.

Asthenosphere   From the Greek *asthenos*—weak. A zone within the Earth's mantle starting between 50 to 100 km below surface and extending to possibly 500 km, between which the rock is sufficiently plastic for overlying plates of lithosphere to move horizontally; or to sink or rise according to the thickening or thinning of the overlying crust.

Basal   Applied to the lowest member of a stratigraphic series.

Basalt   Lava that is relatively mafic (having less than 55% silica) and being dark in colour.

Basement   The crustal layer below sedimentary deposits extending towards the crust-mantle boundary. Generally basement rocks are igneous or metamorphic and Precambrian in age.

Basin   The topographical use of the word indicates a depressed area into which the adjacent land drains. In structural geology it indicates a syncline that is circular or elliptical in plan with the beds dipping inward.

Bathymetric   Relating to measurement of depths especially in oceans, e.g. bathymetric contours.

Bauxite   The principal ore of aluminium composed of one or more aluminium hydroxide minerals. Forms at surface under alternating wet and dry conditions in tropical or semi-tropical climates. See also **Laterite**.

Benioff Zone   Also known as a subduction zone. A dipping zone where one tectonic plate undercuts another, commonly an oceanic plate under a continental plate (see p.11).

Bituminous   Yielding bitumen or solid to semi-solid hydrocarbon, e.g. bituminous shale which is often the source of oil and gas concentrations.

Breccia   Fragmental rock whose components are angular as opposed to conglomerate where the fragments are rounded. May be sedimentary or volcanic or formed by crushing along faults.

Carbonatite   Intrusive carbonate rock associated with alkaline igneous activity, commonly in pipe-shaped bodies. Sometimes contains niobium and uranium in commercial quantities and copper (Palabora mine in South Africa).

Constructive Plate Boundaries   A plate tectonic boundary where new rock is being created by rising lava as the plates spread away from the boundary. Contrasted with "destructive boundaries" where rock at the edge of a tectonic plate is being destroyed and absorbed into the mantle at a subduction zone (see p.12).

Continental Seas   Seas lying within or on a continental shelf region. Contrasted with the deep oceans and seas lying on oceanic crust.

Core   The central part of the Earth, beginning at a depth of about 2,900 km and divisible into an outer core to 4,980 km and an inner core from there to the centre.

Craton   A major structural unit of the Earth's crust, relatively immobile and generally consisting of ancient igneous and/or metamorphic rocks. The word is almost synonymous with "shield".

Crust   As used in the composition of the Earth it is the outer layer overlying the mantle and varying from continental crust (average about 35 km thick) to oceanic crust (about 9 km thick).

Dyke   A tabular body of igneous rock cutting across the structure of adjacent rocks.

Evaporites   The sedimentary units which are deposited from seas or lakes as a result of extensive or total evaporation of the body of water. Examples are gypsum, salt and potash.

Fault   A fracture or fracture zone along which there has been displacement of the sides relative to one another, parallel to the fracture (see p.15).

Felsic   Describes an igneous rock composed of light coloured minerals and containing over 55% silica.

Flood Basalt (or Plateau Basalt)   A term applied to those basaltic lavas that occur as vast accumulations of near-horizontal flows and which erupted in rapid succession over great areas. Have at times flooded the Earth's surface on a regional scale. Usually the product of fissure eruptions.

Garnierite   Name given to various magnesium nickel silicates which tend to form at the base of nickel laterites and make higher grade nickel ore than the latter.

Geosynclines   Major structural and sedimentational units of the crust of the Earth. They consist of elongated basins which become filled with great thicknesses of sediments and often volcanic units. Progressive subsidence of the basin floor occurs under the weight of the accumulated pile of sediments and subsequent deformation results in a fold-mountain chain.

Gneiss (Gneisses)   Banded, coarse-grained rocks formed during high-grade regional metamorphism of rocks of various origins.

Gondwana; Gondwanaland   Southern hemisphere of ancient (pre-250 m.y.) supercontinent including Australia, Antarctica, Africa, South America and India, south of the Ganges River, plus part of Indonesia. With "Laurasia" to the north, Gondwana made up the ancient continent, prior to break-up, that has been named "Pangaea".

Graben   A rock unit down-thrown between two parallel faults; generally long compared to its width.

Grain   The particles or discrete crystals which comprise a rock or sediment.

Granite   A plutonic rock consisting essentially of alkali feldspar and quartz, minor quantities of mica and sometimes hornblende.

Granulite   A rock of high-temperature metamorphic origin having a granular texture and usually consisting of feldspars, pyroxenes and garnets.

Greenstone   A field term applied to altered basic igneous rocks (especially Precambrian) including lavas and tuffs which owe their colour to the presence of chlorite, hornblende and epidote.

Haematite (Hematite) Fe_2O_3. An important ore mineral of iron.

Hinge Zone, Hinge Line   As used here, the line between an area of shallow-water marine sediments and one of deeper-water sediments such as limestones. Often a significant zone in the localisation of lead-zinc deposits.

Horizon   A time-plane in a sedimentary or volcanic series recognisable by some characteristic feature such as specific fossils or lithology.

Hot-Spot   Localised melting region in the mantle below the base of the lithosphere, a few hundred kilometres in diameter and persistent over at least several millions of years and whose existence is inferred from volcanic activity above it. See **Plume** (see p.17).

Igneous   A rock or mineral formed by crystallisation from molten or partially molten material. Contrasted with sedimentary rocks.

Intrusive   An igneous rock unit having, while in the molten state, penetrated into, or between, other rocks but solidified before reaching the surface.

Island Arc   Curved chain of islands like Japan or the Aleutians, generally convex towards the open ocean, bordered by a deep submarine trench and enclosing a deep sea basin (see p.16).

Isotopes   Elements having an identical number of protons in their nuclei, but differing in the number of neutrons. Isotopic elements have the same atomic number, different atomic weights and almost, but not quite, the same chemical properties.

Kimberlite   An ultra-mafic intrusive rock consisting predominantly of olivine and phlogopite mica. Kimberlite is the primary source of diamonds but not all kimberlites are diamondiferous.

Laterite   Residual iron-rich surface deposit resulting from alternating wet and dry seasons in tropical or sub-tropical latitudes.

Laurasia   Hypothetical continent in the northern hemisphere which supposedly broke up at about the end of the Carboniferous Period. Included in this continent were North and Central America, Europe and Asia north of the Himalayas (see p.9).

Left/Right Lateral Movement   On a fault with horizontal movement. When viewed from one side of the fault, if the apparent movement of the opposite side is to the left it is "left lateral" or "sinistral"; if to the right it is "right lateral" or "dextral".

Lithology   Describes the physical characteristic of a rock—colour, composition, grain size.

Lithosphere   The outer rigid part of the Earth's crust consisting of the continental and oceanic crust and the upper part of the mantle. Contrasted with the non-rigid asthenosphere beginning about 50 km below surface.

Loess   A homogeneous unstratified surface layer of silt or fine sand deposited by wind action and derived mainly from desert areas or from barren areas around ice sheets.

Mafic   Describes igneous rock composed predominantly of magnesian silicates but containing under 55% silica. Dark in colour.

Magma   Naturally-occurring mobile rock material generated within the Earth and capable of intrusion (e.g. granites) and extrusion (lavas).

Mantle   A layer of the Earth between crust and core. Upper boundary about 35 km below surface in continents and 10 km below oceans. Lower boundary about 2,900 km below surface (see p.8).

Metamorphism; Metamorphic   Process by which consolidated rocks are altered in composition, texture or internal structure. Pressure, heat and the introduction of new chemical substances are the principal causes.

Mid-Ocean Ridge   A continuous narrow (about 1,500 km) median mountain range extending through the North and South Atlantic Oceans, the Indian Ocean and the South Pacific Ocean. It has a central rift valley up which lava erupts and about which sea floor spreading is believed to take place (see p.6).

Mobile Belt   A portion of the crust of the Earth, usually long compared to its width, that is more mobile, as evidenced by folding and faulting, than the adjoining stable blocks of the crust.

Moraine   Drift (boulders, gravel, sand or clay) deposited by direct glacial action or by water flowing within, or out of, glaciers.

Nappe   A fold in which the axial plane is horizontal or sub-horizontal.

Nodule   A small, more or less rounded body generally harder than the enclosing sediments. **Manganese nodules,** about 3-5 centimetres average diameter, lie in certain areas of the ocean floor and consist of manganese oxide with sufficient content of copper, nickel and cobalt to be of potential economic value.

Ophiolite   A rock of ocean-floor origin, often thrust onto the flanks of continents during geosynclinal mountain-building. Composed of mafic and ultra-mafic rocks both intrusive and volcanic. Increasingly recognised as of economic importance for supplies of nickel, copper, chromite and asbestos.

Orogeny   A period of mountain building particularly by folding but usually also involving faulting, intrusion and metamorphism.

Over-Thrust; Under-Thrust   Either can occur on a low-angle reverse fault with a net slip usually measured in kilometres. If the hanging wall is the active element it is an over-thrust; if the foot wall, an under-thrust. But it is usually impossible to tell which side actively moved.

Pangaea   The name proposed by Alfred Wegener for the "supercontinent" comprising all land masses on Earth. See **Laurasia** and **Gondwana** and also p.9.

Pelagic   A term used to describe the mode of life of those animals which live in the open sea but not on the sea floor.

Peneplain A land surface worn down by erosion to a nearly flat plain.

Plate One of the large segments of the Earth's crust varying in thickness from 50 to 250 km and including a portion of the upper mantle (lithosphere) above the asthenosphere.

Plate Tectonics The movement of plates (see above) relative to those adjoining. (See also **Tectonics**.)

Platform The area of thinner sediments surrounding a craton or shield of older rocks.

Plumes. Generally occur on the surface of a master joint resulting from heat rising from the mantle up through the crust and are thus related to the convection currents active within the asthenosphere. See also **Hot-Spot**.

Plug A volcanic plug (or neck) is the nearly circular vertical feed-channel of a volcano which has been filled with solidified lava and/or pyroclastic material and has subsequently been exposed by erosion of the volcanic cone. Often it is the only remaining trace of an ancient volcanic vent. An intrusive plug is any rather small, more or less vertical, cylindrical mass of igneous rock of plutonic origin.

Porphyry Originally used to describe an intrusive rock containing relatively large crystals set in a finer-grained ground mass. May be applied to lavas having this characteristic. **Porphyry copper**—Disseminated copper minerals in a large body of porphyry but in the commercial sense may be applied to disseminated copper ore on a large scale in any felsic intrusive.

Pyroclastics Fragmental material resulting from a volcanic explosion which has fallen into water and become part of a sediment.

Radiometric Dating The calculation of an age in years of rocks or minerals, by any one of several age determination methods, based on the nuclear decay of natural radioactive elements contained in the material. Hence the use of the term 'radioactive clock'.

Red Beds Term applied to red sedimentary rocks, usually sandstones and shales formed under oxidising conditions so that iron is present in the form of red ferric oxide. **Red bed copper**—sedimentary copper restricted to one or more horizons in red beds and sometimes extending for great distances. The outstanding example is the Kupferschiefer of Germany and Poland.

Reef As used here the term is applied to a rock structure, either mound-like or layered, built by sedentary organisms such as corals and usually enclosed in rock of differing lithology. Since a reef is usually more porous and permeable than the surrounding material it makes a natural site for oil and gas concentrations. The use of the word to denote a mineralised bed or vein, as is the practice in South Africa, is not employed in this Atlas.

Remobilised; Remobilisation The process by which a mineral deposit is dissolved, transported in solution and redeposited in another place, sometimes resulting in an increased grade of ore.

Richter Scale The range of numerical values of earthquake magnitude devised in 1935 by C. F. Richter. The range is from negative values for very low magnitudes to nearly 9 in the case of a major, disastrous, earthquake.

Rift Valley An elongated trough bounded by faults. Approximately synonymous with **Graben**. Most continental parting commences with rifting.

Shield A continental block of the Earth's crust (usually of Precambrian age) that has been relatively stable over a long period of time in contrast to the strong folding of bordering, geosynclinal belts.

Sill A sheet-like body of intrusive igneous rock which approximately conforms to bedding or other structural planes.

Skarn The term is generally reserved for rocks composed mainly of lime-bearing silicates and derived from nearly pure limestones or dolomites into which large amounts of silica, aluminium, iron and magnesium have been introduced from nearby felsic intrusives.

Stock An intrusive mass of plutonic, igneous rock, smaller in size than a batholith (usually less than 100 km^2 in plan) and possessing a roughly circular or elliptical cross section.

Stratiform Composed of layers. **Stratigraphic geology**—the study of stratiform or stratified rocks.

Stromatolites (Stromatoporoidea) Laminated but otherwise structure-less calcareous objects believed to be fossil algae. Found in rocks as old as 3,000 m.y. but also in some modern sediments.

Strike-Slip or Transcurrent Fault A fault in which the net slip is practically in the same direction as the fault strike.

Subduction Descent of one tectonic unit under another. **Subduction zone**—a region along which one crustal plate descends relative to another, e.g. the descent of the Pacific Plate beneath the Andean Plate. Synonymous with Benioff Zone.

Sulphide Zone That part of an orebody or vein not yet oxidised by air or surface water and containing sulphide minerals.

Suture A line or mark of splitting open, or of the joining of two parts, e.g. the line of collision of peninsular India with the main part of southern Asia can be traced over part of its length along the north side of the Himalayas.

Syncline A fold that is (at least originally) concave upward. See also **Anticline**.

Tectonic An adjective used to relate a particular phenomenon to a structural or mountain-building concept. **Plate Tectonics**—a theory of global-scale dynamics involving the movement of many rigid plates of the Earth's crust.

Thrust Sheet; Thrust Slab The block above a thrust fault.

Till Non-sorted, uncompacted sediment deposited by a glacier. **Tillite**—a sedimentary rock composed of cemented till.

Transcurrent Fault Strike-slip fault. Such faults often offset a mid-oceanic ridge.

Trench; Trough A long but narrow depression of the deep sea-floor having steep sides. Such structures tend to occur on the ocean side of a subduction zone or of an island arc.

Tuff A rock formed of compacted volcanic fragments generally less than 4 millimetres in diameter.

Vulcanism Volcanic activity.

Window An area where erosion has penetrated a thrust or a recumbent fold to expose the rocks lying beneath.